LAB STUDIES IN
PHYSICAL GEOLOGY

Second Edition

R.C. FINCH W.L. HEETON, H.W. LEIMER W.C. SMITH

And some rin up hill and down dale, knapping the chucky stanes to pieces wi' hammers, like sae many road makers run daft. They say it is to see how the warld was made.

Sir Walter Scott
St. Ronan's Well — 1824

LAB STUDIES IN
PHYSICAL GEOLOGY

Second Edition

R.C. Finch
W.L. Helton
H.W. Leimer
M.O. Smith

Art by Harry F. Lane

Hunter Textbooks Inc.

COVER PHOTOGRAPH

The cover photograph is of a drawing of Cul-Car-Mac Falls, taken from a plate in Safford's 1869 text, Geology of Tennessee.

Cul-Car-Mac Falls, located downstream from Evins' Mill near Smithville, Tennessee, is formed by a resistant lip of cherty Mississippian limestone (the Ft. Payne Formation) overlying a more easily eroded Devonian-Mississippian shale (the Chattanooga Shale). The cherty limestone forms the massive upper cliff faces, whereas the well-bedded black shale forms the stepped outcrops over which the lower portions of the falls cascade.

James Merrill Safford was State Geologist and Professor of Natural Science in Cumberland University at Lebanon, Tennessee, at the time of publication of his Geology of Tennessee. Safford later became Professor of Geology at Vanderbilt University. His text, with its 550 pages, 8 plates, and large hand-colored, fold-out geologic map of the state, represents near heroic efforts by a pioneer geologist working under the difficult conditions which prevailed in Tennessee during the mid-nineteenth century. War, primitive transportation systems, and the lack of tools and equipment that no modern geologist would consider working without, all failed to deter Safford. By his own estimation his work represents 10,000 miles of geologic travels. That his geologic map and cross sections are still basically sound a century later is sufficient testimonial to the astuteness and perseverance of the man.

©1980, 1978, 1977 by R.C. Finch, W.L. Helton, H.W. Leimer and M.O. Smith

Illustrations, Copyright Hunter Textbooks Incorporated

ISBN 0-89459-214-9

Inquires should be addressed to:

Hli Hunter Textbooks Inc.
823 Reynolda Road
Winston-Salem, North Carolina 27104

Preface

This laboratory manual is designed for use in introductory courses in physical geology where laboratory work is an integral part of the course. The approach is sufficiently general to permit its use as a supplement to any basic textbook.

The manual was written with the idea that it would be as nearly self-contained as possible because many geology departments are finding it difficult and increasingly expensive to maintain multiple copies of the many topographic maps needed in geology labs.

Rock and mineral specimens required are, where possible, of a general nature, to allow maximum flexibility in the selection of lab specimens.

The self-explanatory, self-contained nature of the various exercises should permit the student to do the assignments with a minimum of supervision, using the maps and explanations provided with each exercise.

The manual is divided into two parts, conforming essentially to two quarter-length courses of physical geology. Part I, Exercises 1 through 8, deals mainly with rock and mineral identification, economic geology, and rock deformation and structure. Part II, Exercises 9 through 21, covers such topics as photo interpretation, map reading, landforms and geologic processes, environmental geology, and geologic map interpretation.

MATERIALS NEEDED BY STUDENTS USING THIS MANUAL

1. Geologic Lab Kit:
   Streak plate, glass plate, magnet, pocket knife, colored pencils, 6″ scale, plastic vial for acid, plastic protractor, hand lens.
2. 3H pencil
3. Art gum eraser
4. Graph paper — 20 squares per inch
5. Plastic triangle

Lab Studies in Physical Geology

Contents

PART I: EARTH MATERIALS AND TECTONICS
Exercise 1: Physical Properties of Minerals 1
Exercise 2: Mineral Identification 7
Exercise 3: An Introduction to Rocks 17
Exercise 4: Igneous Rocks 19
Exercise 5: Sedimentary Rocks 23
Exercise 6: Metamorphic Rocks 27
Exercise 7: Economic Geology 31
Exercise 8: Rock Deformation and Structural Geology 33

PART II: LANDFORMS AND GEOLOGIC PROCESSES
Exercise 9: Aerial Photo Interpretation 39
Exercise 10: An Introduction to Map Reading 47
Exercise 11: Methods of Representing Topography 59
Exercise 12: Topographic Profiles 63
Exercise 13: Streams, Stream Systems and Topography 67
Exercise 14: Karst and Groundwater 91
Exercise 15: Topographic Features of Arid Regions 103
Exercise 16: Alpine Glaciation 119
Exercise 17: Continental Glaciation 131
Exercise 18: Shorelines and Coastal Processes 143
Exercise 19: Volcanic Landforms 159
Exercise 20: Man and His Environment 171
Exercise 21: Geologic Maps and Cross Sections 187

ANSWER SHEETS 205

List of Maps

Menan Buttes, Idaho 41
Cookeville East, Tennessee 43
Bright Angel, Arizona 45
Strasburg, Virginia 58
Harrisburg, Pennsylvania 66
Government Springs, Colorado 73
Leavenworth, Kansas-Missouri 75
Campti, Louisiana 77
Natchez, Mississippi-Louisiana 79
Breton Sound, Louisiana 81
Anderson Mesa, Colorado 83
Hollow Springs, Tennessee 85
Plum Grove, Tennessee-Virginia 87
Maverick Spring, Wyoming 89
Mammoth Cave, Kentucky 95
Lost River Area, Indiana 97
Interlachen, Florida 99
Bottomless Lakes, New Mexico 101
Ennis, Montana 107
Furnace Creek, California 109
Antelope Peak, Arizona 111
New Home, Texas 113
Glen Rock NW, Wyoming 115
Ashby, Nebraska 117
McCarthy, Alaska 123
Chief Mountain, Montana 125
Holy Cross, Colorado 127

Hayden Peak, Utah-Wyoming 129
Kaaterskill, New York 135
Palmyra, New York 137
Jackson, Michigan 139
Whitewater, Wisconsin 141
Boothbay, Maine 149
Lynn, Massachusetts 151
Point Reyes, California 153
Beaufort, North Carolina 155
San Clemente Island Central, California 157
Mt. Rainier, Washington 161
Mauna Loa, Hawaii 163
Mt. Dome, California-Oregon 165
Crater Lake National Park
    and Vicinity, Oregon 167
Ship Rock, New Mexico 169
Ajo, Arizona 175
Harriman, Tennessee 177
Riverton, Minnesota 179
Oswego, Kansas 181
Fairborn, Ohio 183
Derouen, Louisiana 185
Flagstaff Peak, Utah 195
Wetterhorn Peak, Colorado 197
Mifflintown, Pennsylvania 199
Danforth, Maine 201
Brisbin, Montana 203

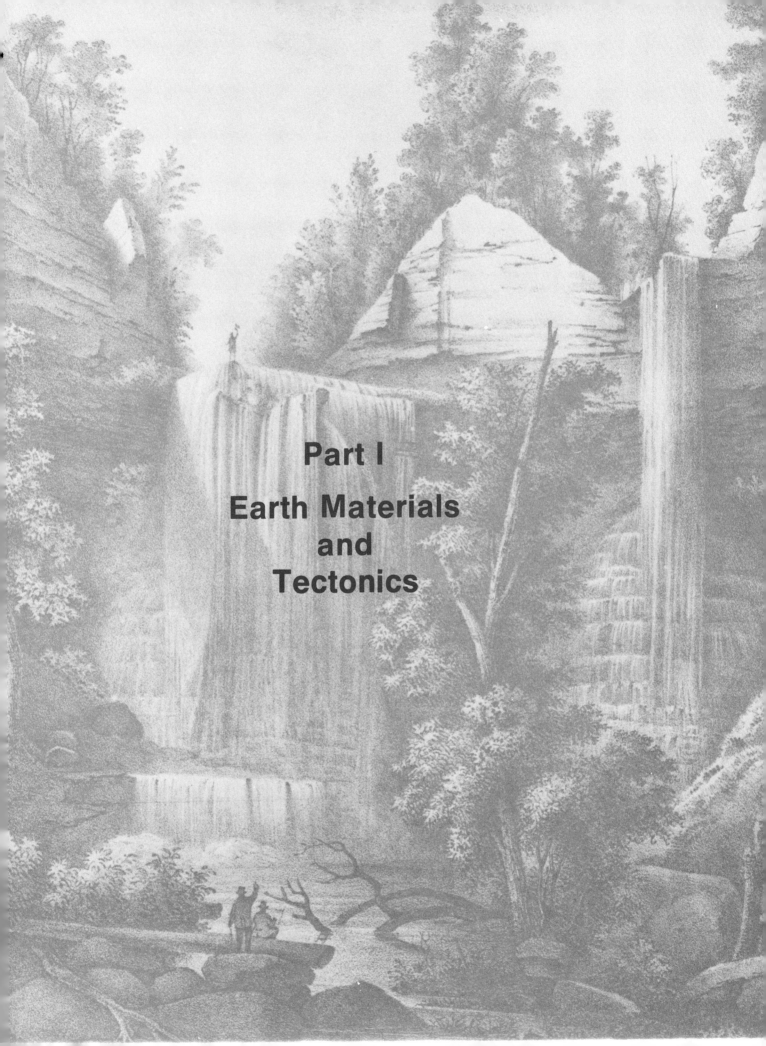

Part I
Earth Materials
and
Tectonics

Exercise 1: Physical Properties of Minerals

The rocks which occur in an area are natural records which can be "read" by geologists to reconstruct the geologic history of the area. Just as one must be able to understand words in order to read and understand sentences in a history book, the geologist must be able to recognize and understand the component parts of rocks in order to interpret the geologic history recorded in the rocks. Rocks (with very few exceptions) are aggregates of **minerals.** Some rocks are composed entirely of one mineral, whereas others contain two or more.

What is a mineral? To be a mineral, a material should fulfill each of the following four criteria:

1. Be naturally occurring. This would eliminate such things as man-made diamonds.

2. Be an inorganic substance.

3. Have a unique or only slightly varying chemical composition. A chemical formula showing the elements present and their ratios can be written. Some minerals consist of only one element, such as graphite (C) or sulfur (S). Most consist of more than one element, such as halite (NaCl), pyrite (FeS_2), or potash feldspar ($KAlSi_3O_8$).

4. Have a definite internal arrangement of the elements present. Atoms in a mineral do not occur in a random manner. They are arranged in an orderly, three-dimensional network similar to bricks in a house. This network is called a lattice.

Some materials fulfill the first three of these criteria but do not have a definite internal arrangement of the elements present. Such materials are said to be **amorphous,** and are called **mineraloids;** for example, limonite.

If atoms were much larger or the magnifying power of the eye much greater, mineral identification would be a simple matter. We could observe directly what elements are present and how they are put together. But instead we must infer the composition and internal arrangement of atoms by observing visible physical properties of the mineral. Each mineral has a characteristic set of physical properties generated by the elements present (composition) and how those elements are put together (internal arrangement). If either the composition or the internal arrangement of atoms changes, the physical properties of the substance also change. The more important physical properties which are helpful in mineral identification follow:

1. **Color.** The color of a mineral is due to the wave length of light reflected by the specimen. Color is the most obvious physical property and can be the least diagnostic. Some minerals occur in only one color. In these cases, such as epidote (green), sulfur (yellow), and azurite (blue), color is diagnostic. Many minerals, however, occur in a variety of colors. Quartz, for example, can be colorless, milky, black, red, yellow, purple, and green. Color should seldom be used as the major identifying physical property.

2. **Streak.** The color of the powdered mineral is referred to as *streak.* The streak is obtained by rubbing the mineral specimen against a piece of white, unglazed porcelain (streak plate). Even though the overall color may vary, the color of the streak is fairly uniform. Both the black, metallic variety of hematite and the red, earthy variety of hematite have a red-brown streak. The color of the mineral may have no relation to the color of its streak. Yellow pyrite has a greenish-black streak. Many minerals with a non-metallic luster have a white or pastel colored streak. In such cases, streak would not be a diagnostic physical property. Minerals which are harder than the streak plate ($> 5\frac{1}{2}$) produce a white streak due to the abrasion of the porcelain and not the mineral, so it is pointless to try to streak these minerals.

3. **Luster.** The appearance of light **reflected from a freshly broken surface** of a mineral is called *luster*. Weathering, handling of the specimens by fellow students, and other factors may affect the luster of a mineral specimen; therefore, be sure to examine a fresh surface. Two broad categories of luster are recognized: (1) metallic and (2) non-metallic. Minerals such as pyrite and galena, which are shiny like the surface of a metal, are said to have a **metallic luster.** Minerals which reflect light like dull metal or metal in need of a polish are said to have a **sub-metallic luster.** Minerals which reflect light like substances other than metal are said to have a **non-metallic luster.** The luster of non-metallic minerals may be further subdivided using descriptive terms as shown in Table 1-1.

TABLE 1-1. TYPES OF NON-METALLIC LUSTERS

Type	Appearance	Mineral Examples
Adamantine	Brilliant glass	Diamond
Vitreous	Like glass	Quartz
Resinous	Like resin	Sphalerite
Waxy	Like wax	Serpentine
Satiny	Like satin	Satinspar Gypsum
Pearly	Like a pearl	Biotite
Greasy	Like bacon grease	Quartz
Dull, Earthy	Like face powder	Kaolinite

4. **Hardness.** The resistance of a mineral to abrasion is called *hardness*. The elements in a mineral are held together by chemical bonds. If the bonds are strong, the resistance to abrasion will be high, and the mineral hard. If the bonds are weak, the resistance to abrasion will be low, and the mineral soft. If bond strength varies with direction, so will hardness. Hardness is determined by comparison with a set of hardness minerals called *Mohs Hardness Scale* (Table 1-2). This scale has been arranged from 1 (the softest) to 10 (the hardest) using minerals representative of each category. If an unknown mineral can be scratched by a mineral of the standard set, then the unknown mineral is softer. To make sure the test mineral has actually been scratched, rub your thumb over the presumed scratch. If a scratch has actually been made on the mineral being tested, it will be visible after the powder has been rubbed away. Common materials with known hardness quickly allow bracketing on Mohs Scale. By using only your thumbnail, a penny, and a pocket knife or glass plate, the hardness can be determined as $< 2\frac{1}{2}$, $2\frac{1}{2}-3\frac{1}{2}$, $3\frac{1}{2}-5\frac{1}{2}$, and $> 5\frac{1}{2}$.

TABLE 1-2. MOHS HARDNESS SCALE

Mohs Number	Mineral	Hardness of Common Materials
10	Diamond	
9	Corundum	
8	Topaz	
7	Quartz	
6	Orthoclase	$5\frac{1}{2}$-6 good quality knife blade
5	Apatite	5-$5\frac{1}{2}$ glass plate
4	Fluorite	
3	Calcite	$3\frac{1}{2}$ penny
2	Gypsum	$2\frac{1}{2}$ thumbnail
1	Talc	

5. **Specific Gravity.** The specific gravity (abbreviated Sp. Gr.) of a mineral is the ratio of its weight to the weight of an equal volume of water. Many common nonmetallic minerals have a specific

gravity between 2.5 and 3.0, that is, they are 2.5 to 3 times heavier than an equal volume of water. Accurate measurement of specific gravity requires specialized equipment. For practical purposes, an estimate of specific gravity can be obtained by comparing the "heft" of the specimen against one of known specific gravity of about the same size. Galena has a specific gravity of 7.5, pyrite about 5, and quartz and orthoclase about 2.7.

6. **Crystal form.** The external shape displayed by minerals which grow in physically unrestricted environments is called *crystal form.* Well-developed crystals are bounded by planar surfaces called **crystal faces.** These faces develop at certain angles to each other, forming the beautiful geometric figures so prized by mineral collectors. The crystal forms that develop are the **external** expression of the definite **internal** arrangement of the atoms which make up the minerals, and therefore certain crystal forms are characteristic of certain mineral species. For example, galena commonly forms crystals shaped like cubes, whereas quartz often forms in hexagonal prisms (and never forms cubic crystals). Some of the more common crystal forms include crystals shaped like cubes, rhombohedrons, rectangular prisms, hexagonal prisms, trigonal prisms, and dodecahedrons. (See Figure 1-1.) There are numerous other crystal forms. Some minerals exhibit more than one crystal form, and some minerals are rarely found in well-developed crystals. Minerals which grow in physically restricted environments, competing for space with other growing minerals, will fail to show good crystal form.

7. **Cleavage.** Bonding in minerals may be of different strengths. If, because of the crystal lattice of the mineral, weak bonds occur in planes and if stress is applied, breakage will occur along these planes. That is known as *cleavage.* Cleavage planes are similar in appearance to crystal faces but are produced by breaking, not by growth. Both are smooth surfaces which reflect light like mirrors. Muscovite and biotite are good examples of minerals with one direction cleavage. They may be separated into sheets much like the individual cards in a deck of cards. The top and the bottom of the sheets are considered the same direction of cleavage since they are parallel. Other numbers of cleavage directions are 2, 3, 4, and 6. (See Figure 1-2.) When 2 or 3 directions of cleavage are present, the angle between the directions must also be given. These angles will approximate either 90° (right angles) or 60° and 120°. All examples of a specific mineral will have a particular cleavage. In some cases though, cleavage may be difficult to see because of small grain size or crystal habit.

8. **Fracture.** Some minerals contain no systematic planes of weak bonding. Breakage occurs along uneven, irregular, or curved surfaces in no definite direction. It is **uneven** fracture when the surface is rough and irregular. The fracture is **conchoidal** when the mineral breaks into smooth and curved, convex-concave faces, like broken glass. A **fibrous** fracture is due to splintering breakage.

9. **Parting.** Parting is defined as breakage along planes of structural weakness which result from imperfections during growth or pressure applied after formation. Parting is similar to cleavage in that breakage is along planes. It is different from cleavage in that not all examples of a specific mineral have parting. Parting is not as common as cleavage. In order to distinguish between cleavage and parting, the beginning student may need additional information.

10. **Tenacity.** The resistance that a mineral offers to breaking, crushing, bending, or tearing is known as *tenacity.* The following terms are used to describe tenacity in minerals:

> **Brittle.** Breaks or powders easily.
>
> **Flexible.** Bends but does **not** resume its original shape when pressure is released — gypsum, chlorite.
>
> **Elastic.** Bends but resumes its original shape when pressure is released — the micas.
>
> **Malleable.** Can be hammered out into thin sheets.
>
> **Sectile.** Can be cut into thin shavings with a knife.
>
> **Ductile.** Can be drawn into wire.

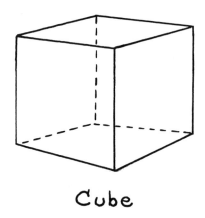

Cube

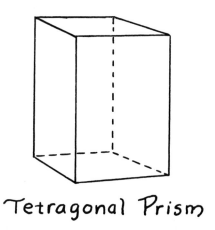

Tetragonal Prism

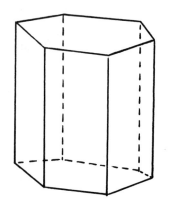

Hexagonal Prism

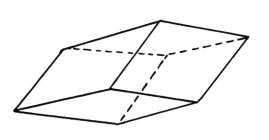

Rhombohedron

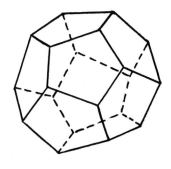

Dodecahedron

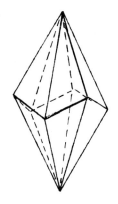

Scalenohedron

Fig. 1-1. Crystal forms

CLEAVAGE IN ONE DIRECTION

CLEAVAGE IN TWO DIRECTIONS AT RIGHT ANGLES

CLEAVAGE IN TWO DIRECTIONS NOT AT RIGHT ANGLES

CLEAVAGE IN THREE DIRECTIONS AT 90°

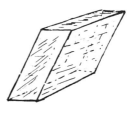

CLEAVAGE IN THREE DIRECTIONS AT 60°

CLEAVAGE IN FOUR DIRECTIONS

CLEAVAGE IN SIX DIRECTIONS

Fig. 1-2. Cleavage

11. **Reaction with hydrochloric acid.** When dilute hydrochloric acid is applied to any carbonate containing mineral, a chemical reaction occurs in which carbon dioxide gas is released. The mineral is said to effervesce or "fizz." Calcite ($CaCO_3$) effervesces rapidly. Other carbonates such as dolomite [$(Ca, Mg)(CO_3)_2$] or siderite ($FeCO_3$) may effervesce slowly or may need to be powdered before effervescence will occur.

12. **Magnetism.** Some minerals, such as magnetite, are attracted by a magnet.

13. **Diaphaneity.** The appearance of the mineral in transmitted light is called *diaphaneity*. If a specimen will transmit both light and an image (i.e., you can see through it) as do window glass or quartz, it is **transparent.** If a specimen will transmit light but not an image as do frosted glass or some varieties of gypsum, it is **translucent.** If neither light nor an image is transmitted even through the thinnest edge, the specimen is **opaque.**

14. **Striations.** Striations are a series of closely-spaced parallel lines or grooves on either crystal faces or cleavage planes. Striations on crystal faces indicate non-uniform growth rates. These striations are surface features only, and do not extend into the crystal. Quartz crystals and pyrite cubes show this type of striations. Striations on the cleavage planes of plagioclase feldspar are due to inversion of the definite internal arrangement of atoms during growth. These striations extend throughout the crystal. White potash feldspar and white plagioclase feldspar have very similar physical properties, but only the plagioclase feldspar has striations.

15. **Other Physical Properties.**

   Taste. Salty — halite.

   Sticks to tongue — kaolinite.

   Odor. Earthy or clayey — kaolinite, bauxite.

   Sound. Crackling noise when held in hand next to ear — sulfur.

   Feel. Greasy or slippery — talc, graphite.

Turn to Page A-1 for questions on material covered in this exercise.

Exercise 2: Mineral Identification

More than 2,500 different minerals have been recognized in nature. Only 40 to 50 are considered important enough to be discussed in this manual. This importance is based on their being (1) constituents of common rocks (Exercises 3, 4, 5, and 6) or (2) economically useful to man (Exercise 7).

The **classification** of minerals used in this laboratory manual is based on chemical composition as shown in Table 2-1.

Mineral identification is simple if a **systematic approach** is used and if the student is familiar with the physical properties as described in Exercise 1. The systematic approach to be used for mineral identification is illustrated in Figure 2-1. By following this technique of recognizing physical properties combined with the organization of the Mineral Identification Tables 2-2 and 2-3, unknown minerals may be identified. The more physical properties that match, the more confident you should be about your identification. Some properties may be variable or difficult to observe. Some properties must be present for accurate identification; e.g., all magnetite is magnetic, all halite tastes salty, all calcite effervesces in dilute HCl.

Procedure

Identify the minerals provided by your instructor. Fill in the Mineral Identification Answer Sheets with physical properties following the technique described in Figure 2-1. Using the Mineral Identification Tables, name the mineral having these properties and record its composition.

Answer sheets for this exercise begin on Page A-5

TABLE 2-1. CLASSIFICATION OF MINERALS BASED ON COMPOSITION.

(These are the minerals listed in the Mineral Identification Tables.)

1. Native Elements (By Itself)

Diamond -	C
Graphite -	C
Sulfur -	S

2. Sulfides (Metal + Sulfur)

Chalcopyrite -	$CuFeS_2$
Galena -	PbS
Pyrite -	FeS_2
Sphalerite -	ZnS

3. Oxides (Metal + Oxygen or Hydroxyl)

Quartz -	SiO_2 (see silicates)
Bauxite -	Al_2O_3 + impurities
Corundum -	Al_2O_3
Hematite -	Fe_2O_3
Limonite -	$FeO(OH) \cdot nH_2O$
Magnetite -	Fe_3O_4
Pyrolusite -	MnO_2

4. Carbonates (Metal + Carbonate)

Calcite -	$CaCO_3$
Dolomite -	$CaMg(CO_3)_2$
Siderite -	$FeCO_3$

5. Halides (Metal + Halogen)

Fluorite -	CaF_2
Halite -	NaCl

6. Sulfates (Metal + Sulfate)

Barite -	$BaSO_4$
Gypsum -	$CaSO_4 \cdot 2H_2O$

7. Phosphates (Metal + Phosphate)

Apatite -	$Ca_5(PO_4)_3(F,Cl)$

8. Silicates (Metal + Silicate)

   Amphibole Group
   Hornblende - Hydrous silicate of
   Ca,Na,Mg,Fe,Ti, & Al
   Beryl - $Be_3Al_2Si_6O_{18}$
   Biotite - Hydrous silicate of
   Al,K,Mg,Fe
   Chlorite - Mg,Fe,Al Silicate
   Garnet - Fe Aluminosilicate
   Kaolinite - $Al_4(Si_4O_{10})(OH)_8$
   Kyanite - Al_2SiO_5
   Montmorillonite - Complex
   Ca,Na,Mg Silicate
   Muscovite - Hydrous Silicate of
   Al,K
   Olivine - $(Mg,Fe)_2SiO_4$
   Plagioclase Feldspar Group

Anorthite	Ca-Rich	⎫
Bytownite	↑	⎬ Na-Ca Aluminosilicate
Labradorite	⏐	
Andesine	↕	
Oligoclase		⎭
Albite	Na-Rich	

   Potash Feldspar Group
   Orthoclase - $KAlSi_3O_8$
   Pyroxene Group
   Augite - Ca,Fe,Mg Aluminosilicate
   Quartz - SiO_2
   Serpentine (Asbestos)
   $Mg_3Si_2O_5(OH)_4$
   Staurolite - Complex Fe
   Aluminosilicate
   Talc - Hydrous Mg - silicate
   Topaz - $Al_2(SiO_4)(OH,F)_2$
   Tourmaline - Complex silicate

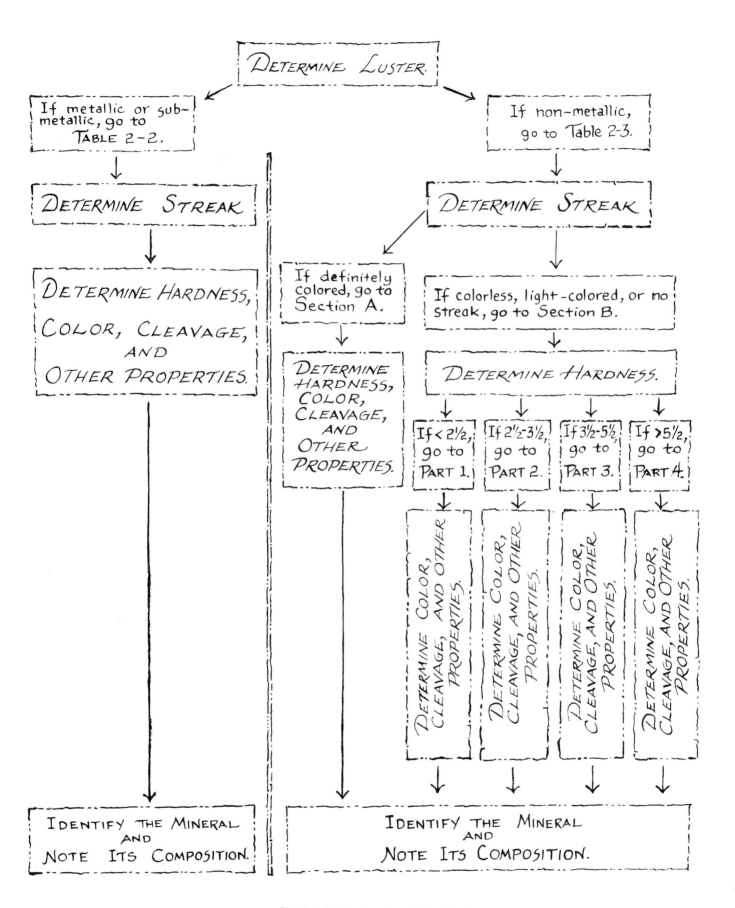

Fig. 2-1. Outline for mineral identification

TABLE 2-2. LUSTER: Metallic or Sub-metallic

STREAK	HARDNESS	COLOR	SP. GR.	REMARKS & USES	NAME AND COMPOSITION
Black	1	Steel gray	2	Soft, marks on paper, greasy feel. 1 direction cleavage. Used in refractory crucibles, lubricants, and pencil "leads."	GRAPHITE C
Iron-black	1-2	Black	4.8	Radiating fibers, granular masses, or dendritic; sooty. An ore of manganese.	PYROLUSITE MnO_2
Yellow brown	1 to 5½	Yellow brown to dark brown to black	3.3 to 4.0	Flattened crystals, massive, reniform, or stalactitic. Secondary mineral in rocks and soils. An ore of iron.	LIMONITE $FeO(OH) \cdot nH_2O$
Red brown to Indian red	1 to 6½	Steel gray to black	4.8 to 5.3	Massive, radiating, micaceous. Crystalline varieties harder than earthy. An ore of iron.	HEMATITE Fe_2O_3
Gray	2½	Gray	7.6	Occurs in cubes; may be massive or granular; heavy; cubic cleavage. The ore of lead.	GALENA PbS
Greenish-black	4	Brass yellow	4.3	Often tarnished purple or gray; yellower and softer than pyrite. An ore of copper.	CHALCOPYRITE $CuFeS_2$
Black	6	Black	5.2	Strongly magnetic. An ore of iron.	MAGNETITE Fe_3O_4
Black to greenish	6 to 6½	Pale brass	5.0	Often in crystals. Massive, granular. Common name: "Fool's gold." Sometimes mined as a source of sulfur.	PYRITE FeS_2

TABLE 2-3. LUSTER: Non-metallic
Section A. Streak definitely colored

STREAK	HARDNESS	COLOR	SP. GR.	REMARKS & USES	NAME AND COMPOSITION
Yellow brown	1 to 5½	Yellow brown to dark brown	3.6 to 4.0	Earthy. Secondary mineral in rocks and soils. An ore of iron.	LIMONITE $FeO(OH) \cdot nH_2O$
Red brown to Indian red	1 to 6½	Red, vermillion	4.8 to 5.3	Earthy; frequently as pigment in rocks. Massive, radiating. Crystalline varieties harder than earthy. An ore of iron.	HEMATITE Fe_2O_3

TABLE 2-3. LUSTER: Non-metallic
Section B. Streak Colorless or Light Colored
PART 1. Hardness: < 2½ (can be scratched with thumbnail)

HARDNESS	CLEAVAGE FRACTURE	COLOR	SP. GR.	REMARKS & USES	NAME AND COMPOSITION
1	Good cleavage in 1 direction	White, green, pink	2.7	Flexible but not elastic; foliated; slick feeling. Used in paints, ceramics, rubber, insecticides, paper.	TALC Hydrous Mg-silicate
1-2	No macroscopic cleavage	White, tan, light to dark gray	2-3	Earthy; clay odor; swelling clay. Used to stop leaks in soils, rocks, dams, and basement walls.	MONTMORILLONITE Complex Ca, Na, Mg Aluminosilicate
1-3	Uneven fracture	Yellow brown to red	2-3	Dull to earthy luster; in rounded grains — pisolitic; not truly a mineral. An ore of aluminum.	BAUXITE Al_2O_3 + impurities
1½-2½	Conchoidal to uneven fracture	Yellow	2.1	Characteristic yellow color; when held in hand close to ear it will be heard to crackle, due to thermal expansion. Used to make sulfuric acid, fertilizers, insecticides, explosives, and medicines.	SULFUR S
2	1 direction perfect	Pale brown, green, yellow	2.8	In foliated masses and scales; transparent, flexible, and elastic sheets. Used as insulating material in electrical apparati and as a fireproofing material.	MUSCOVITE Hydrous Silicate of Al, K
2	No macroscopic cleavage	White, often colored by impurities	2.6	Earthy; clay odor; non-swelling clay; sticks to tongue. Used in refractories, china, pottery, and as a filler in paper.	KAOLINITE $Al_4(Si_4O_{10})(OH)_8$
2	1 direction, perfect, 2 directions, good	Colorless, white, gray, gray-brown, pink reddish	2.3	As crystals and broad cleavage flakes (Selenite); as compact masses showing no cleavage (Alabaster); as fibers with satiny luster (Satinspar). Used to make plaster of paris and wallboard.	GYPSUM $CaSO_4 \cdot 2H_2O$
2-2½	1 direction	Dark green to green-black	2.7	Flexible sheets.	CHLORITE Mg, Fe, Al Silicate
2-3	Wavy, uneven fracture	Green and white	2.5	Platy or fibrous, waxy luster when massive. Used as insulating material against heat and electricity.	SERPENTINE (Asbestos) $Mg_3Si_2O_5(OH)_4$

TABLE 2-3. LUSTER: Non-metallic
Section B. Streak Colorless or Light Colored
PART 2. Hardness: 2½ - 3½ (cannot be scratched with thumbnail;
can be scratched with penny)

HARDNESS	CLEAVAGE FRACTURE	COLOR	SP. GR.	REMARKS & USES	NAME AND COMPOSITION
2-3	Wavy, uneven fracture	Green and white	2.5	Platy or fibrous, waxy luster when massive. Used as insulating material against heat and electricity.	SERPENTINE (Asbestos) $Mg_3Si_2O_5(OH)_4$
2½	3 directions, perfect, cubic	White when pure; may be red, blue, pink, etc.	2.1 to 2.3	In granular cleavable masses or cubic crystals. Soluble in water; salty taste. Common salt. A source of sodium and chlorine for sodium compounds and hydrochloric acid; used to salt highways in winter; a seasoning.	HALITE $NaCl$
2½ to 3	1 direction, perfect	Dark brown, green to black	3.0	As irregular foliated masses and scales; transparent, flexible, and elastic sheets.	BIOTITE Hydrous Silicate of Al, K, Mg, Fe
3	3 directions, perfect, rhombic	White or colorless, but may be pink, blue, brown, etc.	2.7	Crystals in many forms. Occurs as large granular masses (limestone or marble) and fine granular or fibrous masses in which cleavage not prominent; compact masses. Effervesces in cold, dilute HCl. Used in the manufacture of cement; crushed stone; agricultural lime.	CALCITE $CaCO_3$
3 to 3½	1 direction, perfect. 2 directions, good	White or gray	4.5	Crystals usually tabular; very heavy for a nonmetallic. Used to give weight to drilling muds in order to prevent "blow-outs" of oil and gas wells.	BARITE $BaSO_4$

TABLE 2-3. LUSTER: Non-metallic
Section B. Streak Colorless or Light Colored
PART 3. Hardness: 3½-5½ (cannot be scratched with penny;
can be scratched with knife)

HARDNESS	CLEAVAGE FRACTURE	COLOR	SP. GR.	REMARKS & USES	NAME AND COMPOSITION
3½ to 4	3 directions, perfect, rhombic	White, pink, brown, gray, etc.	2.9	Usually harder than penny. As crystals with curved faces (twisted rhombs). As granular masses (dolomitic marble, dolostone). Effervesces in cold, dilute HCl only if powdered. Used as a building and decorative stone.	DOLOMITE $CaMg(CO_3)_2$
3½ to 4	Perfect cleavage in 6 directions	Yellow to brown, black, reddish brown	4.0	Resinous luster, usually massive. All six cleavages rarely seen at same time. An ore of zinc.	SPHALERITE ZnS
3½ to 4	3 directions, perfect, rhombic	Light to dark brown; maroon	4.0	As crystals with curved faces. Usually cleavable; sometimes granular masses. Effervesces in dilute HCl only if powdered. Minor ore of iron.	SIDERITE $FeCO_3$
4	Good in 4 directions, octahedral	Purple, green to yellow, colorless	3.2	Well-formed cubic crystals, also massive. Used as a flux in steel making, and in the production of hydrofluoric acid.	FLUORITE CaF_2
5	Poor cleavage, 1 direction	Green to brown	3.2	Glassy, massive and granular. Used as a source of phosphate for fertilizers.	APATITE $Ca_5(PO_4)_3(F,Cl)$
5 to 6	2 directions, good, at approx. 60° and 120°	Green to black	3.0 to 3.3	Crystals slender, fibrous. Commonly in cleavage fragments or granular masses.	AMPHIBOLE GROUP (Hornblende) Hydrous silicate of Ca, Na, Mg, Fe, Ti, and Al
5 to 7	Good in 1 direction	Blue to green	3.6	In bladed aggregates. Used in the manufacture of spark plugs and other highly refractory porcelains.	KYANITE Al_2SiO_5
5½ to 6	2 directions, poor to fair, at approx. 90°	Green to black	3.1 to 3.5	Crystals "stubby" with rectangular cross section. Commonly in granular, crystalline masses.	PYROXENE GROUP (Augite) Alumino-silicate of Ca, Mg, and Fe

HARDNESS	CLEAVAGE FRACTURE	COLOR	SP. GR.	REMARKS & USES	NAME AND COMPOSITION
5 to 6	2 directions, good, at approx. 60° and 120°	Green to black	3.0 to 3.3	Crystals slender, fibrous. Commonly in cleavage fragments or granular masses.	AMPHIBOLE GROUP (Hornblende) Hydrous silicate of Ca, Na, Mg, Fe, Ti, and Al
5 to 7	Good in 1 direction	Blue to green	3.6	In bladed aggregates. Used in the manufacture of spark plugs and other highly refractory porcelains.	KYANITE Al_2SiO_5
5½ to 6	2 directions, poor to fair, at approx. 90°	Green to black	3.1 to 3.5	Crystals "stubby" with rectangular cross section. Commonly in granular, crystalline masses.	PYROXENE GROUP (Augite) Alumino-silicate of Ca, Mg, and Fe
6	2 directions, good at right angles	Colorless, white, pink, red, gray, green, etc.	2.5 to 2.6	As cleavable masses or irregular grains in rocks. As crystals in pegmatites and some igneous bodies.	POTASH FELDSPAR GROUP (Orthoclase) $KAlSi_3O_8$
6	2 directions, good, at approx. 90°	Colorless, white, various shades of gray	2.6 to 2.8	In cleavable masses or irregular grains. Striations common.	PLAGIOCLASE FELDSPAR GROUP $NaAlSi_3O_8$ - $CaAl_2Si_2O_8$
6½	Uneven fracture	Red to brown	4.3	Usually in 12 or 24 sided crystals, also massive. Some samples may exhibit parting. Used as an abrasive, and as a gemstone.	GARNET (Almandite) $Fe_3Al_2(SiO_4)_3$
6½ to 7	Conchoidal fracture	Olive green to yellow green	3.3 to 3.4	Usually as disseminated grains in mafic igneous rocks; as granular masses having saccharoidal texture (looks like grains of sugar). Mined for refractory sand used in the casting industry.	OLIVINE $(Mg, Fe)_2SiO_4$
7	Conchoidal fracture	Colorless or white when pure, but may be any color	2.6	As crystals with hexagonal cross section often with striations on prism faces. As crystalline masses, granular aggregates, irregular grains etc. Vitreous or greasy luster. Varieties — *Milky:* white and opaque, *Smoky:* gray to black, *Rose:* pink, *Amethyst:* violet. Used as a gemstone, a flux, a filler, and an abrasive.	QUARTZ SiO_2
7	Conchoidal fracture	Various colors	2.6	Varieties — *Agate:* massive to banded, *Flint:* dark color, *Chert:* light color, white to gray, *Jasper:* red, *Opal:* milk-white, yellow, green, red, etc., waxy luster, *Chalcedony:* brown to gray, fibrous.	MICROCRYSTAL-LINE QUARTZ SiO_2

TABLE 2-3. LUSTER: Non-metallic
Section B. PART 4, Continued

HARDNESS	CLEAVAGE FRACTURE	COLOR	SP. GR.	REMARKS & USES	NAME AND COMPOSITION
7 to 7½	Cleavage not prominent	Varied; black common	3.2	Usually in trigonal prismatic crystals; striations prominent. Used as a gemstone.	TOURMALINE Complex silicate
7 to 7½	Cleavage not prominent	Red-brown to brownish-black	3.7	Cross-shaped twin crystals common ("Fairy crosses").	STAUROLITE Complex Fe Aluminosilicate
8	Imperfect cleavage	Green to yellow	2.7	Hexagonal, prismatic crystals. Used as a gemstone, and as a source of beryllium for metal alloys.	BERYL $Be_3Al_2Si_6O_{18}$
8	1 direction, poor	Colorless, pink, yellow	3.5	Prismatic crystals, crystalline masses, or granular. Used as a gemstone.	TOPAZ $Al_2(SiO_4)(OH,F)_2$
9	Parting, no true cleavage	Brown, pink, ruby-red	4.0	Barrel-shaped crystals; hexagonal prisms; basal parting common. Used as an abrasive, and as a gemstone (red — ruby, and blue — sapphire).	CORUNDUM Al_2O_3
10	4 directions	Colorless, pale yellow	3.5	Adamantine luster, uncut crystals have a characteristic greasy appearance. Used as an abrasive, and as a gemstone.	DIAMOND C

Exercise 3: An Introduction to Rocks

A rock is a naturally occurring mass of inorganic or organic material that forms a significant part of the earth's crust. Rocks are the materials of the earth's crust which a geologist studies in order to reconstruct the geologic history of the earth. In studying rocks, geologists have recognized three major rock families, each major family of rocks being distinguished from the others by its mode of origin. The three rock families are interrelated to each other as shown in Figure 3-1, the Rock Cycle.

Those rocks formed from the cooling of a high temperature melt called a *magma* (or *lava*) are termed **igneous** (Latin, *fire-formed*). A magma may originate by melting mantle material from the interior of the earth or by melting previously formed crustal rocks. It may cool beneath the earth's surface or on it. Igneous rocks, sediments, sedimentary rocks, or metamorphic rocks may be subjected to weathering, transportation, deposition, and lithification to form **sedimentary** (Latin, *to settle*) rocks. When igneous rocks, sedimentary rocks, and even metamorphic rocks are subjected to heat, pressure, and chemically active fluids (an environment different from that under which they were formed) **metamorphic** (Greek, *changed in form*) rocks will be generated. When a metamorphic rock is remelted, a magma is formed and the cycle starts again.

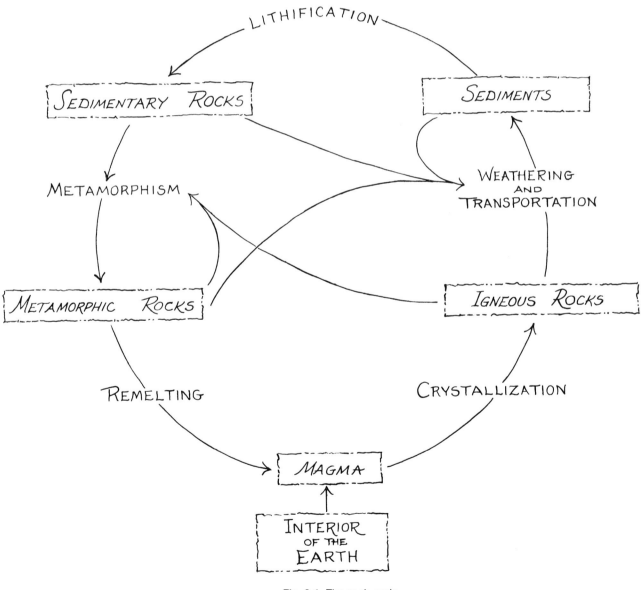

Fig. 3-1. The rock cycle

The environment of formation of a rock strongly influences its mineral constituents or **composition.** Some rocks are composed of only one mineral. Others are composed of more than one mineral. Still other types, such as coal, are composed mainly of organic material with a very small amount of true minerals. The **common rock-forming minerals** are listed in Table 3-1. Note that some minerals are characteristic of only one rock family, such as olivine in igneous and kyanite in metamorphic. Some are characteristic of two rock families, such as calcite in sedimentary and metamorphic. Still others are characteristic of all three, such as quartz. In general, igneous rocks contain minerals which are stable at high temperatures, sedimentary rocks contain minerals which are stable at low temperatures, and metamorphic rocks contain minerals which are stable at a temperature between that of igneous and sedimentary rocks.

The environment of formation of a rock also strongly influences the shape and size relationship, or **texture,** of the mineral constituents. Igneous rocks formed by crystallization of a magma contain interlocking mineral grains. The rock is usually hard and dense. Some sedimentary rocks may contain fragments of previously formed rock material which has been weathered, transported, abraded, deposited, and then cemented together, whereas others are composed of interlocking crystals formed by chemical precipitation from water. The latter type of sedimentary rock is often monomineralic (one mineral). Most metamorphic changes take place within the earth's crust. As a result of the pressures, many platy or prismatic minerals become oriented to each other, producing a layering or **foliation.** Furthermore, recrystallization may convert many small crystals into fewer larger crystals.

Mineral composition and texture are the two characteristics used in rock classification.

Answer sheets for this exercise begin on Page A-11.

TABLE 3-1. COMMON ROCK-FORMING MINERALS

Igneous Rocks	Sedimentary Rocks	Metamorphic Rocks

Plagioclase Feldspar
  Anorthite   Ca-Rich
  Bytownite
  Labradorite
  Andesine
  Oligoclase
  Albite   Na-Rich
	Calcite
	
	Dolomite
	
	Gypsum
		Garnet
		
		Staurolite
		
		Kyanite

Potash Feldspar Orthoclase		Calcite
	Halite	

Quartz	Clay Minerals
	  Kaolinite
	  Montmorillonite
		Micas
		
		Talc

Micas
  Muscovite
  Biotite
	Iron Oxides
	  Hematite
	  Limonite
		Chlorite

Pyroxenes
  Augite
	Quartz
	  Opal
		Serpentine

Amphiboles
  Hornblende
	Chert
	  Chalcedony
		Graphite
		
		Quartz

Olivine	Orthoclase	
	Also many of the igneous rock-forming minerals occur in this group, as detrital minerals.	Also many of the igneous and sedimentary rock-forming minerals may be formed by metamorphic processes.

Exercise 4: Igneous Rocks

Igneous (Latin, *fire-formed*) rocks form when magma cools and solidifies. The characteristic properties, that is, texture and composition, of the resulting igneous rock depend upon the following four factors:

1. **The composition of the original magma.** Magmas containing abundant iron, magnesium, and calcium (the ferromagnesian elements) are called *mafic* or *basaltic* magmas. Calcium-rich plagioclase feldspars and pyroxenes with minor amounts of amphiboles and olivine are common minerals which form from a basaltic magma. Magmas that are unusually rich in the ferromagnesian elements are called *ultramafic* magmas. Olivine and pyroxenes with minor amounts of calcium-rich plagioclase feldspars are common minerals which form from an ultramafic magma. Mafic and ultramafic igneous rocks are usually dark-colored because they contain dark-colored minerals. Magmas that are rich in silicon, aluminum, potassium, and sodium are called *felsic* or *granitic* magmas. Quartz, potash feldspar, and sodium-rich plagioclase feldspar are common minerals which form from a granitic magma. Felsic or granitic igneous rocks are usually light-colored because they contain light-colored minerals. A magma of a composition intermediate between granitic and basaltic will form quartz, sodium-rich plagioclase feldspar with some potash feldspar, and amphiboles. This rock will be intermediate in color between light and dark because it contains a mixture of light- and dark-colored minerals. Certain elements are needed in a magma before minerals containing these elements can form. For example, no potash feldspar can form unless potassium is present in the magma. Thus the composition of the original magma affects the composition of the final igneous rock.

2. **Whether the first-formed mineral crystals remain in the magma or are removed from the magma.** In the 1920's N.L. Bowen devised Bowen's Reaction Series (BRS) (Figure 4-1) through experiments involving man-made magmas which were allowed to cool under controlled laboratory conditions. The data from these tests show that, as a magma cools, different minerals crystallize at different temperatures, forming two mineral series. Minerals in the **continuous series** belong to the plagioclase group, all having the same crystal structure and differing only in composition. Minerals in the **discontinuous series** differ both in crystal structure and in composition. As a basaltic magma cools, the first minerals to crystallize are olivine and Ca-rich plagioclase, then pyroxene. At this stage two things may happen to change the composition of the magma. These first-formed minerals may sink to the bottom of the magma chamber leaving the magma remaining in the upper part of the magma chamber poorer in Ca, Mg and Fe than it was originally. Furthermore, the magma may be squeezed through cracks in the surrounding rocks, causing the early formed crystals to be "filtered out." In either case the composition of the remaining magma is altered. A third possibility is that the early formed minerals remain with the main magma body, and as the temperature falls some of these minerals will react chemically with the magma to dissolve and reform as minerals lower down in the BRS. It may be seen by the BRS that certain mineral associations may be expected in a magma which has cooled to a certain point. For example, amphiboles, biotite, and Na-plagioclase form in the same temperature range. If a magma is rapidly chilled at a certain point in its cooling history, this may prevent crystals of late forming minerals from developing. Thus the mineral composition of the resulting rock may be partly determined by the stage of the BRS reached before final cooling. The BRS may also be used to show mineral antipathies: it would be extremely rare to find quartz and olivine associated together in the same igneous rock.

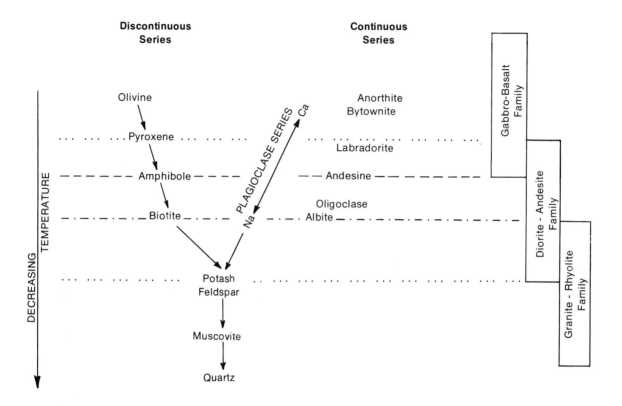

Fig. 4-1. Bowen's Reaction Series

3. **Rate of cooling of magma.** Magmas which solidify below the surface of the earth are said to be **intrusive.** Magmas which reach the earth's surface (termed **lavas**) are said to be **extrusive.** The size of the crystals is controlled by the rate of cooling. Most intrusive and some extrusive rocks have cooled at a uniform slow rate. The crystals have time to form and grow. This produces an equigranular igneous rock with crystals large enough to be seen with the unaided eye. This texture is termed coarse-grained or **phaneritic.** Most extrusive and some intrusive rocks have cooled at a uniform fast rate. The crystals have time to form but not enough to grow in size. This produces an equigranular igneous rock with crystals too small to be seen with the unaided eye. This texture is termed fine-grained or **aphanitic.**

Two different rates of cooling are reflected in an igneous rock by two different crystal sizes. Minerals formed earlier during a slow rate of cooling are larger than those formed during a later, faster rate of cooling. The larger crystals are called **phenocrysts.** The smaller mineral constituents are referred to as the **groundmass** or **matrix.** When two distinct grain sizes are present, the textural term **porphyritic** or **porphyry** is used. If the phenocrysts are 5% to 25% of the rock, the adjective *porphyritic* is used before the composition. If the phenocrysts are > 25% of the rock, the composition is placed before the noun *porphyry.* See Table 4-1 for an example of the usages of these terms.

Some extrusives cool at a super fast rate. The constituents in the magma are "frozen" so rapidly that no organization into crystals can occur. A **noncrystalline solid,** similar to a glass, is formed. This texture is termed **glassy.** Obsidian is a volcanic glass. If the igneous rock has phenocrysts but a glassy groundmass, the rock name **vitrophyre** is used.

% Phenocrysts	Groundmass	Texture	Rock Name
0-5%	Phaneritic	Phaneritic	Granite
5-25%	Phaneritic	Porphyritic Phaneritic	**Porphyritic** Granite
> 25%	Phaneritic	Phaneritic Porphyritic	Granite **Porphyry**
0-5%	Aphanitic	Aphanitic	Rhyolite
5-25%	Aphanitic	Porphyritic Aphanitic	**Porphyritic** Rhyolite
> 25%	Aphanitic	Aphanitic Porphyritic	Rhyolite **Porphyry**

Solidified material may be blown out of a volcano to form rocks containing fragments or broken pieces instead of interlocking grains. This type of texture is called **fragmental.** This category includes **tuff** and **volcanic breccia.**

Thus the rate of cooling affects the texture of the final igneous rock and, in some cases, its composition.

4. **Volatile content.** Magmas always contain some volatile materials (gases) kept in solution by the pressure. The most important volatile constituent of magmas is water vapor. Other volatiles are fluorine, chlorine, hydrogen, nitrogen oxides, sulfur oxides, and carbon dioxide. The greater the volatile content, the more fluid the magma and lava, which flow easier. The greater the volatile content the lower the temperature at which a magma will crystallize; i.e., the magma will stay molten at a temperature lower than usual. The greater the volatile content, the larger the crystals that will form due to the increased mobility of elements in the highly fluid melt, as well as to the greater length of crystallization time. **Pegmatites** are extremely coarse-grained igneous rocks formed by volatile-rich magmas. As the pressure is reduced on a magma or lava, the volatiles may separate from the liquid. Escaping gas bubbles trapped during rapid crystallization are called **vesicules.** The rock looks like Swiss cheese or a sponge. **Pumice** is a light-colored, vesicular, glassy igneous rock; **scoria** is the dark-colored equivalent. Volatiles are also necessary components for the formation of certain minerals. If these components are not present, for example, the $(OH)^-$, hydroxyl group, the amphibole hornblende will not form.

Thus the volatile content affects both the texture and the composition of the final igneous rocks.

As shown in Figure 4-2, both compositional and textural properties are used in the classification of igneous rocks. Rock types falling within horizontal rows have the same texture. Rock types falling within the same vertical column have the same composition. By determining the texture and composition of a specimen, the intersection of a row and a column is determined and the rock name is "pigeonholed."

Procedure

Classify each of the igneous rock specimens furnished by your instructor. Use the igneous rock identification answer sheets provided. First, determine texture. Second, determine the color: light, intermediate, dark. Finally, determine what minerals are present, if possible, and their approximate percentage. Combine the textural and compositional characteristics and compare these to the Igneous Rock Classification Chart (Figure 4-2). Name the rock.

Answer sheets for this exercise begin on Page A-15.

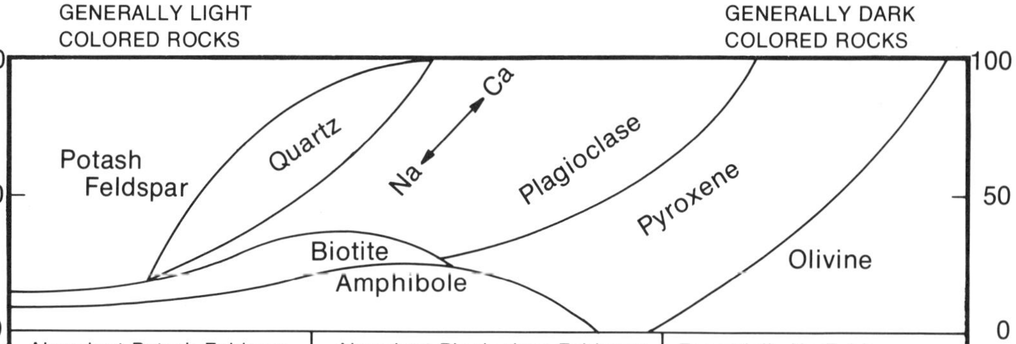

Classification diagram. Top portion — COMPOSITION / TEXTURE axes (0–100) with mineral distribution curves labeled: Potash Feldspar, Quartz, Na–Ca, Plagioclase, Pyroxene, Olivine, Biotite, Amphibole.

			Abundant Potash Feldspar		Abundant Plagioclase Feldspar		Essentially No Feldspar	
			No Quartz	Quartz	Na-Plagioclase Plus Amphibole	Ca-Plagioclase Plus Pyroxene	Pyroxene Plus Olivine	Olivine
Phaneritic (Coarse)	Non-Porphyritic		Syenite	**Granite**	Diorite	**Gabbro**	Peridotite	Dunite
	Porphyritic	5–25% Phenocrysts	Porphyritic Syenite	Porphyritic **Granite**	Porphyritic Diorite	Porphyritic Gabbro	Porphyritic Peridotite	Porphyritic Dunite
		>25% Phenocrysts	Syenite Porphyry	**Granite** Porphyry	Diorite Porphyry	Gabbro Porphyry	Peridotite Porphyry	Dunite Porphyry
Aphanitic (Fine)	Non-Porphyritic		Trachyte	**Rhyolite**	Andesite	**Basalt**		
				Felsite				
	Porphyritic	5–25% Phenocrysts	Porphyritic Trachyte	Porphyritic **Rhyolite**	Porphyritic Andesite	Porphyritic Basalt		
		>25% Phenocrysts	Trachyte Porphyry	**Rhyolite** Porphyry	Andesite Porphyry	Basalt Porphyry		
Glassy	Non-Porphyritic		**Obsidian, Pumice,** Scoria					
	Porphyritic		Vitrophyre					
Fragmental			Tuff, Volcanic Breccia Kimberlite					

Igneous rock types shown in **bold print** are the most common varieties.

Fig. 4-2. Classification of igneous rocks

Exercise 5: Sedimentary Rocks

Although more than 90% of the rocks of the outer 10 km of the earth are igneous, more than 75% of the rocks covering the earth's surface are sedimentary (Latin, *to settle*). This relationship is similar to that of the white fruit and red peel of an apple. This is what would be expected since sedimentary rocks are formed by surficial geologic processes.

The following four steps are involved in the formation of sedimentary rocks:

1. **Weathering.** When igneous rocks, sediments, sedimentary rocks, or metamorphic rocks (see Rock Cycle in Exercise 3) are exposed at the earth's surface to an environment different from that in which they were formed, they become unstable. The compositional and textural changes in response to the new environment are called *weathering.* There are two types of weathering:

   Mechanical. Mechanical weathering is a disintegration process. Smaller particles are made out of larger particles. No new minerals are formed, no old minerals disappear. Mechanical weathering may occur because of thermal expansion and contraction, frost action, abrasion during transportation, plant and animal activity, or exfoliation. The greatest contribution of mechanical weathering is to increase the surface area of the mineral grains which in turn makes chemical weathering more effective.

   Chemical. Chemical weathering is a decomposition and recomposition process. Old minerals disappear, new minerals form. The rate of chemical weathering is controlled by surface area of mineral grains, amount of rainfall, temperature, slope of the land surface, and the composition of the original parent rock. Bowen's Reaction Series (Figure 4-1) can be used to show the influence of the composition on the chemical weathering of an igneous rock. Olivine, pyroxene, and Ca-rich plagioclase feldspar are the first minerals to form out of the magma. The high temperature environment under which they were formed is far removed from the conditions at the earth's surface today. These minerals would be most unstable and would chemically weather rapidly. Quartz and potash feldspar are the last minerals to form out of the magma. The lower temperature environment under which they were formed is not as far removed from the conditions at the earth's surface today. These minerals would be more stable and would chemically weather slowly. The first formed minerals on Bowen's Reaction Series chemically weather rapidly, while the last formed weather slowly. All other factors being equal, a gabbro will chemically weather faster than a granite.

2. **Transportation.** Material generated by weathering may be transported by water, wind, ice, or gravity. Most weathered mineral grains are transported as particles either in suspension or as bed load by water or wind currents. If the velocity of the transporting medium exceeds the **terminal velocity** of the particles as determined by its size, shape, and specific gravity, the particle is transported in **suspension.** It is free of the bottom environment. If the velocity of the transporting medium does not exceed the terminal velocity of the particle, it may be transported as **bed load** either by saltation, rolling, or sliding. The particle is never totally free of the bottom environment. Chemical weathering also produces some materials dissolved in solution (in water) which are transported in that manner. During transport, further mechanical and chemical weathering occur. Angular and different size grains are produced at the source. As transportation occurs, abrasion rounds the grains and a sizing or sorting occurs. If the sediment still contains angular grains of different sizes, it

TABLE 5-1. STEPS IN THE FORMATION OF SEDIMENTARY ROCKS

STEP 1 WEATHERING		STEP 2 TRANSPORTATION		STEP 3 DEPOSITION		STEP 4 LITHIFICATION		PRODUCT ROCK TEXTURE

Mechanical	→	As Particles
		1. In Suspension
		2. In Bed Load
		  (a) saltation
		  (b) rolling
		  (c) sliding
		3. Carried by ice
		4. Moved downslope by gravity
			→	By Settling	→	Cementation
						
						or
						
						Compaction
						
						or
						
						Desiccation
							→	Detrital:
								Aggregates of individual grains

Chemical	→	In Solution	→	By Evaporation or By Chemical Reaction or By Organic Activity	→	Crystallization	→	Chemical or Bio-chemical: Interlocking crystals, ooids, or biochemical grains.

TABLE 5-2. CLASSIFICATION OF SEDIMENTARY ROCKS.

TEXTURE	PARTICLE TYPE		COMPOSITION	COMMENTS			ROCK NAME	
Detrital (Grains can be seen with the unaided eye)	INDIVIDUAL GRAINS	> 2 mm	Any rock type with quartz, chert, or quartzite predominant	Rounded particles			Conglomerate	SANDSTONES
				Angular particles			Breccia	
		$\frac{1}{16}$ – 2 mm (sand sized)	Rock fragments, mica, clay, quartz	"Dirty" looking, dark colored; sandpapery feel			Graywacke	
			Quartz, with at least 25% potash feldspar	Often reddish because of potash feldspar; sandpapery feel			Arkose	
			Quartz with minor accessory minerals	White, tan, brown; sandpapery feel			Quartz Sandstone	
Fine (Particles cannot be seen with the unaided eye)	(Detrital grains or interlocking crystals)		Quartz and clay minerals	Gritty, 1/16-1/256 mm; some grains can be seen with hand lens			Siltstone	
			Predominantly clay minerals	Smooth, < 1/256 mm; non-laminated			Claystone, Mudstone	
			Predominantly clay minerals	Smooth, < 1/256 mm; laminated			Shale	
			Calcite ($CaCO_3$)	Consolidated; fizzes rapidly with dilute HCl			Micrite	
			Calcite ($CaCO_3$)	Powdery; shells of microscopic animals; fizzes with HCl			Chalk	
			Dolomite $(Ca,Mg)(CO_3)_2$	Consolidated; fizzes with dilute HCl when powdered			Dolostone	
			Chalcedony (SiO_2)	Light colored, hardness of 7			Chert	
			Chalcedony (SiO_2)	Dark colored, hardness of 7			Flint	
			Carbonaceous Material	Plant Remains	Brown, soft, porous		Peat	
					Brown		Lignite	
					Black, sooty, blocky		Bituminous Coal	
Chemical or Biochemical (Crystals can be seen with the unaided eye)	INTERLOCKING CRYSTALS, OOIDS, or BIOCHEMICAL GRAINS		Calcite ($CaCO_3$)	Fizzes rapidly with dilute HCl	Medium to Coarse Grained		Crystalline Limestone	
					Abundant fossils, consolidated		Fossiliferous Limestone	
					Oolites, like fish eggs		Oolitic Limestone	
					Banded		Travertine	
					Fossils & fossil fragments loosely cemented		Coquina	
			Dolomite $(Ca,Mg)(CO_3)_2$	Fizzes with dilute HCl when powdered			Dolostone	
			Halite (NaCl)	Evaporite, tastes salty			Rock Salt	
			Gypsum ($CaSO_4 \cdot 2H_2O$)	Evaporite, hardness of 2			Rock Gypsum	

is referred to as being **texturally immature.** If, on the other hand, sufficient abrasion and sorting have occurred during transportation, the sediment is referred to as being **texturally mature. Compositionally immature** sediments still contain unstable minerals. **Compositionally mature** sediments contain stable minerals. The degree of maturity is a measure of the time and distance involved in transportation from the source area to the place of deposition.

The combined effects of weathering and transportation are commonly known as **erosion.**

3. **Deposition.** Material which is transported as particles either in suspension or as bed load is deposited by **settling** when the velocity of the transporting medium is slowed or stopped. Material which is dissolved will not be deposited by settling; deposition will only occur by precipitation when the solution is evaporated or a chemical reaction forms some sort of insoluble material or the material is extracted by organic activity.

4. **Lithification.** Lithification is the conversion of unconsolidated material to that which is consolidated. This is achieved by **cementation, compaction,** or **desiccation** (drying out) of material which has settled. Lithification by **crystallization** occurs during deposition of the material transported in solution.

Table 5-1 summarizes the steps in the formation of sedimentary rocks. These steps lead to a genetic classification of sedimentary rocks into two major groups. Grains which are produced by mechanical and chemical weathering, transported as particles, deposited by settling, and lithified by cementation, compaction, and/or desiccation fall into one group — **detrital** (also called *clastic* or *fragmental*). Materials which are produced by chemical weathering, transported in solution, deposited by evaporation or a chemical reaction or organic activity, and lithified during deposition by crystallization fall into the other group — **chemical** or **biochemical.**

The classification of sedimentary rocks is shown in Table 5-2. Three major textural categories are recognized: (1) those composed of aggregates of detrital grains; (2) those composed of interlocking crystals (similar in appearance to phaneritic igneous rocks); and (3) those composed of particles too small to determine if they are individual grains or interlocking crystals. As the students work with this third category, they will begin to see that some of these sedimentary rocks fall into category (1) and others into category (2).

Procedure:

Describe the texture, particle size, mineral composition and other properties of the sedimentary rocks furnished by your instructor. Fill in this data on the sedimentary rock identification answer sheets. Compare this information with the Sedimentary Rock Classification Chart and name the rock.

BEWARE: Some sandstones are cemented with calcite. They will effervesce when acid is added. Make sure you identify the grains in addition to the cement. (Remember: quartz grains will scratch glass.)

Answer sheets for this exercise begin on Page A-17.

Exercise 6: Metamorphic Rocks

Metamorphic (Greek, *changed in form*) rocks are formed when pre-existing rocks are changed markedly in texture and/or composition by heat, pressure, and chemically active fluids. These pre-existing rocks may have been either igneous rocks, sedimentary rocks, or other metamorphic rocks (see Rock Cycle, Exercise 3).

There are four objectives in studying metamorphic rocks.

1. **Identification and classification of the rock.** Using both texture and composition (as with igneous and sedimentary rocks), we can give the rock a specific rock name (Table 6-1).

2. **Identification of original rock.** Noting the changes that have taken place during metamorphism may enable us to tell what the original rock was before metamorphism occurred (Table 6-2).

3. **Tell what type of metamorphism occurred.** There are two main types of metamorphism. **Contact metamorphism** occurs near the contacts of a cooling magma. Heat and chemically active fluids are important agents of metamorphism while pressure plays a subordinate role. **Regional metamorphism** occurs in orogenic (mountain building) belts or at tectonic plate boundaries. Heat and directed stresses are important agents of metamorphism while chemically active fluids play a subordinate role.

4. **Tell the grade of metamorphism.** The grade of metamorphism refers to the intensity with which the agents of metamorphism were active. Low-grade metamorphic rocks are characterized by the presence of chlorite and biotite, medium-grade by garnet and staurolite, and high-grade by kyanite and sillimanite.

There are three agents of metamorphism.

1. **Heat.** The increase in temperature may be due to nearby igneous activity, radioactive decay, the normal increase in temperature with increasing depth or perhaps frictional heat produced during diastrophism (deformation of the rock by crustal movements).

2. **Pressure.** The increase in pressure may be due to the weight of overlying rocks, tectonic forces within the earth, or perhaps the forceful injection of a magma.

3. **Chemically active fluids.** Water is the most common chemically active fluid. This water may be **juvenile water** associated with magmatic activity, **connate water** trapped in the sediments when they formed, **meteoric water** which has found its way into the crust through fractures and faults, or **water of formation** which is an integral part of the composition of a particular mineral, especially the clays, micas, amphiboles, and gypsum.

As already mentioned, the word *metamorphic* means changed in form. What types of changes occur because of the agents of metamorphism? These changes can be categorized as follows:

1. **Shearing.** The increase in pressure can shear, fracture, or pulverize the rock, such as along faults.

2. **Growth of new minerals.** Minerals with the same composition as the original minerals but with a different internal arrangement may be the more stable forms at higher temperatures and pressures. Chemically active fluids may introduce new elements, generating new minerals.

TABLE 6-1. CLASSIFICATION OF METAMORPHIC ROCKS

	TEXTURE	PARTICLE SIZE	COMPOSITION	COMMENTS	ROCK NAME
ORIENTED GRAINS	Foliation	Fine grained, minerals not visible	Clay minerals, micas	Dense	Slate
			Clay minerals, micas	Satiny luster	Phyllite
	Foliation or Lineation	Medium to coarse grained, minerals visible	Muscovite, biotite, chlorite, talc, garnet, kyanite, staurolite, quartz, ferromagnesian minerals.	Rock name is preceded by diagnostic minerals such as garnet mica schist, kyanite biotite schist, hornblende schist	Schist
	Banding		Feldspars, quartz, micas, ferromagnesian minerals.	Banding due to alternation of light and dark minerals	Gneiss
NON-ORIENTED GRAINS		Medium to coarse grained, minerals visible	Calcite (CaCO$_3$)	Hardness of 3; fizzes rapidly with dilute HCl	Marble
			Dolomite (Ca,Mg) (CO$_3$)$_2$	Fizzes with dilute HCl only when powdered	Dolomitic Marble
			Quartz (SiO$_2$)	Hardness of 7; breaks across grains	Quartzite
			Amphiboles	Generally black; prismatic crystals with 2 directions of cleavage at 120°	Amphibolite
			Anything that could be a conglomerate	Breaks across grains as well as around them	Metaconglomerate
		Fine grained, minerals not visible	Clay minerals, micas	Dense, dark colored	Hornfels
			Carbonaceous material	Black, shiny, conchoidal fracture	Anthracite Coal

3. **Recrystallization.** Due to pressure, some mineral grains dissolve, while others grow, resulting in a rock which is coarser grained than the original. A marble is always coarser grained than the limestone from which it was formed. The pore space and water content of the rock are reduced, while the rock density is increased. Crystals with a preferred orientation grow while the others dissolve or fail to grow. Flaky minerals develop which give some metamorphic rocks a characteristic texture of oriented grains.

4. **Differential melting.** Sometimes the temperature increases enough to melt some of the low-melting point minerals but not enough to melt the whole rock, thus forming a magma. Once these rocks cool, they take on a spotted appearance.

5. **Baking.** If the heat is not enough to melt part of the rock, it still can get hot enough to acquire a "baked" appearance.

6. **Segregation of minerals.** During recrystallization and differential melting, some mineral grains migrate together to form groups or concentrations of the mineral; gneissic banding is due to this type of segregation.

7. **Orientation of mineral grains.** Mineral grains which are platy, prismatic, or sheet-like become oriented perpendicular to the pressure applied. Parallelism of sheet-like minerals such as muscovite, chlorite, and talc is termed **foliation.** Parallelism of prismatic minerals such as hornblende is termed **lineation.** An alternation of colors due to an alternation of mineral layers is termed **banding.**

TABLE 6-2. PRODUCTS OF METAMORPHISM

| | Original Rock | LOW GRADE | | MEDIUM GRADE | | HIGH GRADE |
| | | | INDEX MINERALS | | | |
		Chlorite Zone	Biotite Zone	Garnet Zone	Staurolite Zone	Sillimanite Zone
IGNEOUS ROCKS	Rhyolite	Rhyolite	Biotite schist or gneiss	Biotite schist or gneiss	Biotite schist or gneiss	Biotite schist or gneiss
	Granite	Granite	Granite gneiss	Granite gneiss	Granite gneiss	Granite gneiss
	Basalt	Chlorite-epidote-albite schist	Chlorite-epidote-albite schist	Albite-epidote amphibolite	Amphibolite	Amphibolite
SEDIMENTARY ROCKS	Limestone and dolostone	Limestone and dolostone	Marble	Marble	Marble	Marble
	Quartz sandstone	Quartzite	Quartzite	Quartzite	Quartzite	Quartzite
	Clayey sandstone	Micaceous sandstone	Quartz-mica schist	Quartz-mica garnet schist or gneiss	Quartz-mica garnet schist or gneiss	Quartz-mica garnet schist or gneiss
	Shale	Slate	Phyllite	Biotite-garnet phyllite	Biotite-garnet staurolite schist	Sillimanite schist or gneiss

The metamorphic rock classification chart is shown in Table 6-1. The classification is descriptive as to texture (oriented vs. non-oriented grains) and not genetic (contact vs. regional). Most rock names are sufficient by themselves, but the rock name schist must be preceded by the diagnostic minerals present, such as *staurolite muscovite* schist.

Procedure:

Examine and describe the metamorphic rocks provided by your instructor. Use the metamorphic rock identification answer sheets. Identify each rock using the Metamorphic Rock Classification Chart (Table 6-1). Also list the possible original rock, using Table 6-2.

Answer sheets for this exercise begin on Page A-19.

Exercise 7: Economic Geology

The objective of this exercise is to study the occurrence and characteristics of some of the important economic mineral resources. In the **mining industry** an **ore** is a concentrated accumulation of mineral matter from which one or more **metals** can be extracted at a profit. Ore consists of a mixture of **ore minerals** and valueless materials called **gangue**. In addition to ores, mineral resources include nonmetallic **industrial minerals** and **fossil fuels.** Economically valuable minerals may occur as fine **disseminations** throughout a large rock body, as mineralized **veins** cutting across another rock body, or as large **replacement bodies** which literally take the place of pre-existing rock bodies. They may also occur as **secondary products of weathering** (as newly formed minerals or as residual minerals). The main processes by which economic deposits are formed are summarized in this exercise.

Igneous Processes

The early-formed, high temperature ore deposits result within magmas by the process of **magmatic segregation.** Magma includes not only elements which crystallize into common rock-forming minerals, but also heavy metallic elements and light watery and gaseous material. During the early stages of crystallization, mineral grains of greater density sink through the fluid magma and become concentrated in layers or bands near the base of an igneous body. Elements commonly concentrated in this manner include iron (magnetite), titanium (ilmenite), chromium (chromite), nickel (pentlandite), tungsten (wolframite), and native platinum.

With continued decrease in temperature and pressure, the more common rock-forming minerals such as hornblende, feldspar, quartz, and mica crystallize to form granite and numerous other igneous rocks. Since the ore-forming metals are not major constituents of these rocks, those metals remaining in the magma after magmatic segregation will be concentrated out of the gaseous and watery fluids during the late stages of crystallization. These form an important class of ore deposits known as **hydrothermal.** These solutions leave a magma and travel upward along fractures, cracks, or fissures in the surrounding country rock. At lower temperature and pressure, the hydrothermal solutions are no longer able to hold metals in solution; therefore, minerals precipitate and are deposited in fractures to produce ore **veins.** Where many veins are closely spaced, they constitute a **lode.**

During late stages of magmatic crystallization, the melt consists of low-temperature silicates, water, volatiles, and usually some exotic elements. Such a melt crystallizes slowly, forming coarse-grained igneous rocks called **pegmatites.** Rare and unusual minerals like topaz, tourmaline, lepidolite, and beryl are often deposited with the rock minerals potash feldspar, mica, and quartz.

When an intrusive magma of silicic or intermediate composition comes in contact with certain rocks, particularly carbonates, high-temperature fluids from the magma may completely alter or metamorphose these rocks. During this metamorphism these fluids may introduce ore minerals into the surrounding rock, usually forming replacement ore bodies near the igneous contact. Some important deposits of iron, tungsten, copper, lead, and zinc have been formed in this manner and are called **contact metasomatic deposits.**

Sedimentary Processes

Sediment deposition is the principal source of non-metallic mineral deposits and of large quantities of sedimentary iron ore. Three classes of sedimentary deposits are recognized.

Organic. The most important resources of this group are coal, oil shale, petroleum, and natural gas. Coal is an accumulation of partially altered plant material formed *in situ*; it occurs as sedimentary beds within other sedimentary beds. Buried accumulations of microscopic organic matter are the source materials of petroleum and natural gas.

Chemical. Easily dissolved or soluble materials like gypsum, salt, potash, and borax are commonly carried by streams to the sea without deposition. In arid regions, water carrying these substances are locally trapped in lakes where evaporation may cause salts to be concentrated and eventually precipitated. Temporary lakes in arid regions frequently dry up completely, and salts contained in the water are left as a salt crust on the dry lake floor. Notable examples include the salt flats around Great Salt Lake, Utah, and the borax deposits of Death Valley, California.

Sedimentary Rock Deposits. Iron and manganese in the form of oxides are common sedimentary rocks, particularly iron, occurring as the mineral hematite. The iron at Rockwood, Tennessee, and Birmingham, Alabama, and other deposits scattered throughout the Appalachian region are sedimentary deposits occurring within sedimentary rocks of Early Silurian age. Limestone and sandstone are valuable as building material. Clay, which occurs in many different varieties, is the basic material used in the brick, pottery, and ceramic industries. Less commonly, greensand, bentonite, chalk, and phosphate occur as valuable sedimentary deposits.

Weathering Processes

Residual Concentrations. Some rocks contain minerals which are more resistant to weathering than other minerals in the same rock. Under conditions of strong chemical weathering (leaching), the minerals of no value may be dissolved out and carried away if they are soluble. This leaves a concentration of the insoluble, more resistant valuable compounds. Iron oxides, bauxite (aluminum oxides and hydroxides), manganese oxides, nickel silicates, and clays, all respond favorably to this type of concentration.

Mechanical Concentration. Insoluble heavy minerals derived from weathering of rock are concentrated through agents of transportation, particularly streams and waves. Heavy minerals which are also durable enough to withstand the rigors of water transport are carried downstream with the sand and gravel. Because of their greater density, these minerals are deposited on the inside of meanders or anywhere the water velocity has been greatly reduced. Such concentrations of valuable minerals in sands and gravel are called **placers.** Native gold is one of the minerals extensively mined from placer deposits. Diamonds are concentrated in placer deposits, as are several semiprecious gemstones. Other important placer concentrations include cassiterite (tin), native platinum, ilmenite (titanium), and chromite (chromium).

Ground-Water Processes

Secondary ores result from the effects of water moving downward from the surface. Certain ore minerals, especially copper minerals occurring with pyrite, may be dissolved above the water table by downward percolating water and redeposited as new ore minerals (copper sulfides) in more concentrated forms at or near the water table. Low grade, non-economical deposits are upgraded and made commercial through **secondary enrichment.** Thus **supergene** enrichment is effective in concentrating dispersed metals, and many of the large copper deposits of the western United States owe their economic success to this process.

Ground water, especially potable water utilized for human needs, constitutes a valuable mineral resource. **Connate water** (fossil seawater) trapped in sediments often contains appreciable amounts of salt, iodine, or bromine in solution which may be treated as recoverable mineral resources.

Metamorphic Processes

In addition to the contact metamorphism and metasomatism already described, metamorphism of rocks by heat, pressure, or fluids may change minerals of no value into new minerals or rocks of economic importance. Ultramafic igneous rocks and dolomitic limestones may be changed into asbestos or talc. Metamorphism may also change the form of a rock without altering its composition, thus producing a valuable rock, as in the case of the change of limestone to marble and of shale to slate.

Turn to Page A-21 for questions on material covered in this exercise.

Exercise 8: Rock Deformation and Structural Geology

Rock Deformation

Everyday experience tells us that rocks are hard, rigid, and unyielding. It requires considerable force, such as a strong hammer blow, to deform an ordinary rock, and that rock, when it deforms, is likely to shatter suddenly into many separate pieces. Nevertheless, even the casual observer must notice that almost all rock units wherever exposed are broken by numerous fractures. Furthermore, some rock units have been tilted, bent, and folded. And some of these folded units have been distorted without coming apart, demonstrating that, under the proper conditions, rocks may behave as if they were made of soft putty instead of hard, rigid material. The geologist wants to know how rocks may be deformed, how they behave during deformation, and, ultimately, the cause of deformation. These are not easy subjects to investigate, especially the question of cause.

Under the low temperatures and pressures found in the upper parts of the earth's crust, rocks tend to behave as **brittle substances.** Such a substance undergoes **elastic deformation** when a **stress**[1] is exerted on it, until the **elastic limit** is exceeded; then it fails by **rupturing.** What is elastic deformation? The classic example is a spring, which changes shape (deforms, or strains) by compressing or stretching when it is pushed upon or pulled on, then snaps back to its original shape when the push or pull is relieved. But if the spring is over-strained, it will snap into pieces. Rocks are like this under the conditions we normally deal with them. Hit a rock with a hammer and it compresses very slightly under the hammer; then as the force of the blow dies away, the rock springs back into its original shape. (Why do you suppose the hammer bounces back off the rock surface?) But if hit hard enough, the elastic limit is exceeded and the rock ruptures under the hammer blow, fracturing and separating into pieces.

Under the high temperatures and pressures found deeper in the earth's crust, rocks may behave as **ductile substances.** Such a substance, when stressed, deforms elastically at first, but as the elastic limit is exceeded it begins to undergo **plastic** (permanent) **deformation.** As in elastic deformation, the ultimate result may be rupture if deformation becomes excessive. A good example of a plastic substance is modeling clay: apply a stress and it first deforms elastically (only very slightly), then it takes on another shape that is permanent; it does not snap back to its original shape when the stress is relieved. Deep within the earth's crust rocks are mashed, folded, stretched, and otherwise distorted plastically; the results of these deformations can be seen when the rocks are later uplifted and exposed by erosion at the surface.

Other factors are also important in determining how a rock may deform. The **rock type** is important: granite is more brittle than most sandstone, and sandstone is more brittle than most limestone, etc. Another important factor is the **strain rate.** Laboratory experiments and observations of rocks in natural occurrences show that the faster a rock is forced to deform, the less likely it is to deform plastically and the more likely it is to rupture brittly. If a rock is strained very slowly it may deform plastically even under low temperatures and pressures. Old-fashioned horizontal marble-slab tombstones just a few hundred years old may develop noticeable sags with only the force of gravity pulling on them. Considering the many millions of years that geologic processes may operate, it is not surprising that major plastic deformation may be the result of slow strain induced by relatively low stresses.

---

1. **Stress** is force per unit area, or pressure (but as used here does not refer to the overall confining pressure due to depth of burial in the earth's crust).

Actually, the mechanical characteristics of rocks may be thought of as being very similar to "silly putty." This material stretches plastically if pulled slowly, can be rolled into a ball which has a good elastic bounce when dropped on a hard surface, and shatters into pieces if bounced too hard or struck by a hammer. Rocks behave similarly; all that is needed to achieve plastic deformation of rocks is low strain rates, or high temperatures and pressures.

The two graphs (Figure 8-1) compare elastic and plastic behavior. The graph on the left might represent the deformation of a rock under surface conditions by a hammer blow. The graph on the right might be a rock buried relatively deep in the earth's crust, being deformed by tectonic stresses.

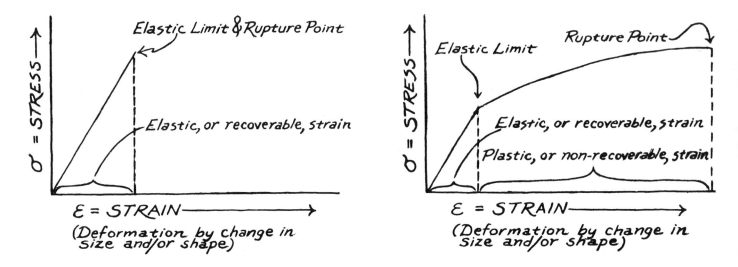

Fig. 8-1. Comparison of elastic and plastic behavior of rocks

Strike and Dip

Before beginning the discussion of geologic structures, strike and dip should be explained. The standard geologic method of describing the orientation of a joint, a fault plane, a rock layer, or any other planar feature is to record its strike and dip. **Strike** is the compass direction of a horizontal line drawn across the planar surface being described. **Dip** is the direction and amount of slope exhibited by the surface. Dip is always measured at right angles to strike inasmuch as this is the direction of maximum slope. The dip is reported by recording the compass direction of the downward slope of the planar surface and also the downward angle from the horizontal. Figure 8-2 illustrates strike and dip, using the case of a planar bed of rock which crops out along the edge of a pool of water.

The surface of the pool of water naturally is a horizontal surface and therefore defines a horizontal line along the bed surface: this line represents the strike of the bed. The dip is measured at right angles to the strike line, as shown. This bed strikes northeast and dips 30 degrees to the southeast. This might be reported in the geologist's field notebook as "strike N45E, dip 30° SE," and shown on his map with a symbol that looks like this: ⟋ 30
This description would be understood without equivocation by other geologists.

Note that special symbols are required if the planar surface is perfectly horizontal or perfectly vertical. In the case of a horizontal plane, any line drawn on the plane is a horizontal line and therefore any line is a strike line; the dip is zero. In the case of a vertical plane the dip has no

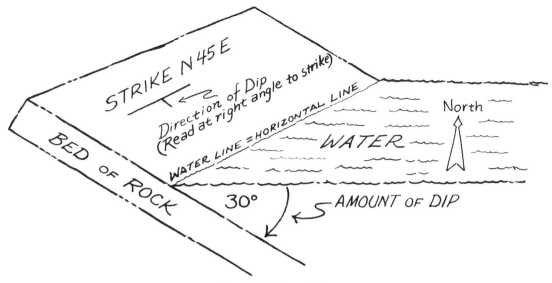

Fig. 8-2. Strike and dip

compass direction, being instead 90° from the horizontal, straight down. The special map symbols are illustrated below.

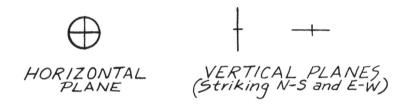

HORIZONTAL
PLANE

VERTICAL PLANES
(Striking N-S and E-W)

Structural Geology

The branch of geology that studies fractures, folds, dislocations and other rock deformations produced by mechanical stresses is called **structural geology.** There are three major classes of rock **structures:** joints, faults, and folds.

Joints

Joints are the most common type of rock structure, being ubiquitous in rocks of the upper portions of the earth's crust. A **joint** is a fracture or break cutting across a rock unit, along which little or no movement has occurred (Figure 8-3A). Joints are the result of brittle failure of the rock due either to tension or shear stresses. Joints are present in all rock units exposed at the surface of the earth whether strongly deformed or nearly undisturbed. Evidently the stresses which cause the rock to fracture may be of many different origins, ranging from mountain building stresses to diurnal earth tides, and including some non-tectonic stresses such as shrinking due to cooling or drying. Next time you go for a drive, notice how all the rocks exposed in the road cuts are broken by various fractures, most of which are joints.

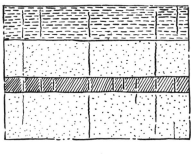

A. Joints

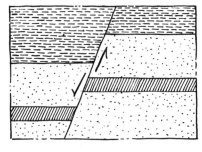

B. Faulting by brittle failure

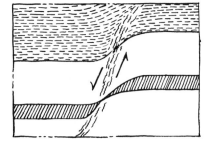

C. Faulting by plastic deformation

Fig. 8-3. Cross sections of jointed and faulted rocks

Faults

A **fault** is a fracture or zone along which noticeable movement has occurred, causing once continuous rocks to be dislocated. A fault due to brittle failure may be a simple fracture, called the **fault plane,** in which case it may look like a joint, except that the rocks on one side of the fracture differ from those on the other side because of movement along the fault (Figure 8-3B). Many faults are not single fractures but networks of fractures which form a **fault zone.** Other faults, especially those occurring deeper in the earth's crust, may result from plastic deformation, in which case there may be no fractures but a zone along which the rocks have been sheared plastically until dislocated (Figure 8-3C).

There are three basic classes of faults: normal, reverse, and strike-slip. A **normal fault** results from a pulling apart of the rocks, so normal faults are considered evidence of **tension** in the faulted block. Figure 8-4A illustrates a normal fault in sedimentary rocks. Note that the block overlying the fault plane, called the **hanging wall,** has moved down the dip of the fault plane relative to the underlying block, called the **footwall.** The sense of movement is shown by half-arrows placed along the fault plane.

On a **reverse fault** the relative displacement is precisely opposite that of a normal fault, and this requires that the rocks be pushed together instead of pulled apart. Thus reverse faults are considered evidence of **compression** in the faulted block. Figure 8-4B illustrates a reverse fault. Note that the hanging wall has been pushed up the dip of the fault plane relative to the footwall. In the case of a reverse fault with a fault plane that lies at a low angle (less than 45° dip), the horizontal displacement will exceed the vertical displacement. This type of reverse fault is called a **thrust fault** (Figure 8-4C).

In both normal and reverse faults the sense of movement along the fault is predominantly up or down the fault plane. But in many faults the movement may be mostly horizontal along the strike of the fault plane. Such a fault is known as a **strike-slip fault.** These faults may be classified as **left-lateral** or **right-lateral,** depending on the direction of relative offset. Figure 8-4D shows right-lateral movement. Sketch for yourself a left-lateral strike-slip fault.

When the fault movement is neither horizontally along the fault plane strike nor up or down the dip of the fault plane, but at some odd angle between these two directions, the movement is called **oblique slip.**

All of these faults are defined according to the **relative movement** between the blocks on opposite sides of the fault. Ask your lab instructor to explain what is meant by "relative movement."

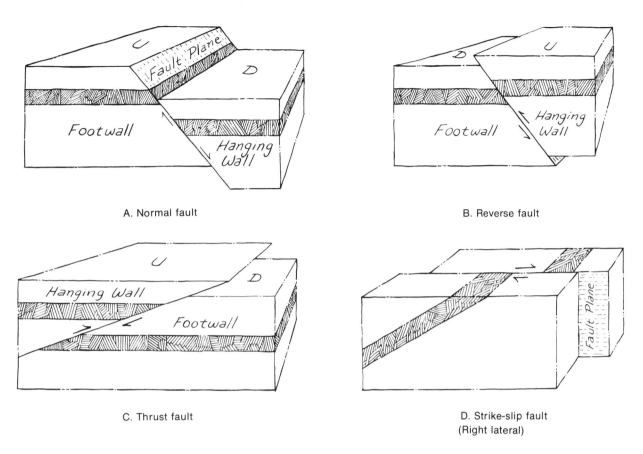

A. Normal fault

B. Reverse fault

C. Thrust fault

D. Strike-slip fault
(Right lateral)

Fig. 8-4. Basic types of faults

Folds

Folds involve bending of the rocks without rupture or with only partial rupture. Therefore folds are primarily the result of plastic deformation. Although folds occur in all kinds of rocks, igneous, metamorphic and sedimentary, they are most easily seen and classified in rocks with originally horizontal layers. Therefore the accompanying figures are of folds developed in sedimentary rocks.

There are two general types of folds: upfolds known as **anticlines,** and downfolds known as **synclines.** These folds are generally considerably longer than they are wide, consisting of two **limbs** which dip away from the **fold hinge** in the case of an anticline, and toward the fold hinge in the case of a syncline. Figure 8-5A is a block diagram of strata which have been folded into an anticline; Figure 8-5B shows a syncline. In both cases the blocks have been eroded so that the upper surface of the block exposes the internal structure of the folds. Observe the patterns made by the rock units and note the strike and dip symbols. Note that in an eroded anticline the **oldest** beds are exposed in the center of the fold; in an eroded syncline the **youngest** beds are found in the center of the structure.

If an upfold is more nearly round than elongate, it is called a **dome;** a nearly equant downfold is called a **structural basin.** See Figure 8-5, C and D.

Answer sheets for this exercise begin on Page A-23.

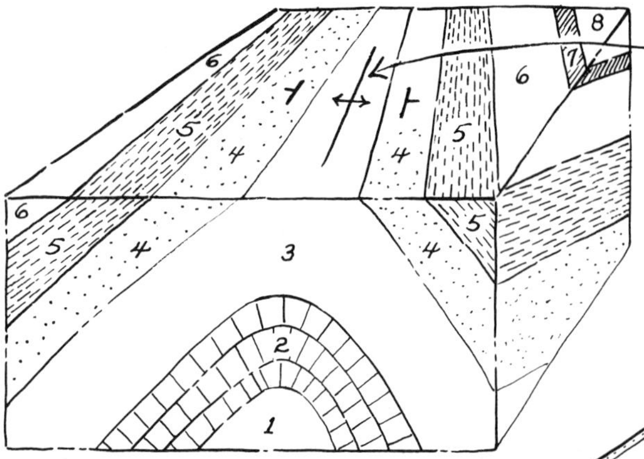

Fig. 8-5a: ANTICLINE
Map Symbol:

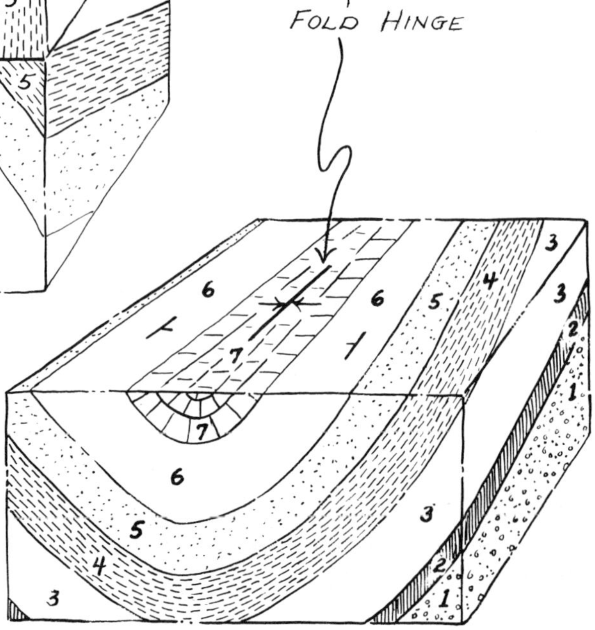

Fig. 8-5b: SYNCLINE
Map Symbol:

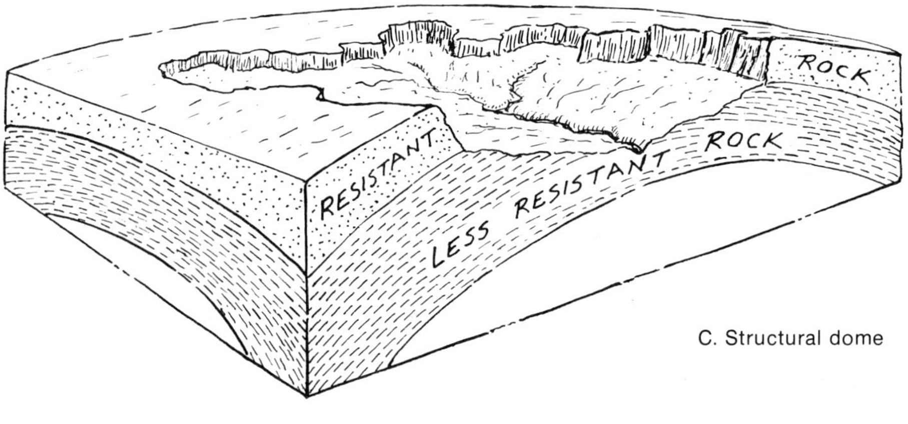

C. Structural dome

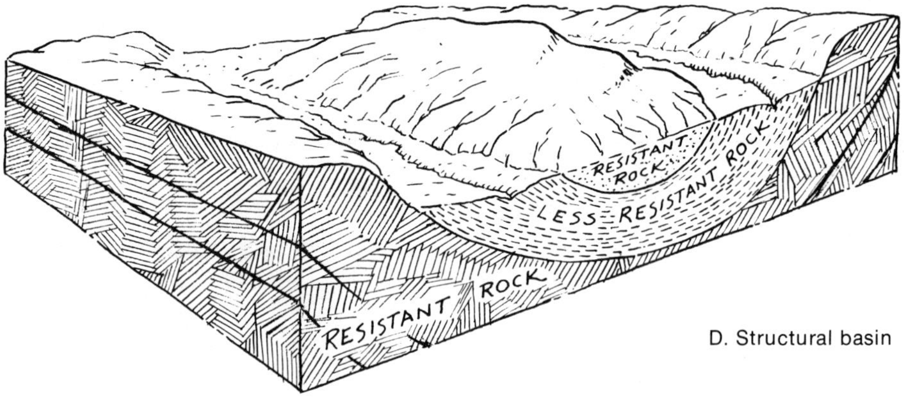

D. Structural basin

Fig. 8-5. Basic types of folds

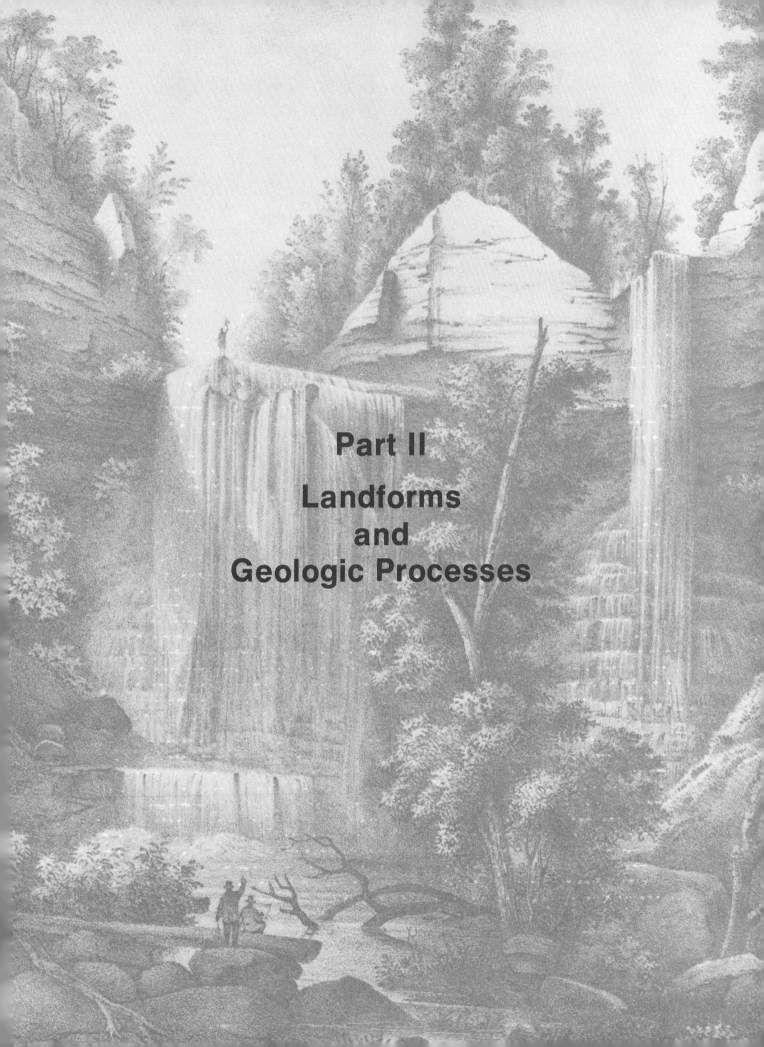

Part II

Landforms
and
Geologic Processes

Exercise 9: Aerial Photo Interpretation

Topographic maps and aerial photographs are the basic tools used in the study and interpretation of landforms and the geologic processes which shape them. The purpose of this exercise is to familiarize the student with the use of stereopair aerial photographs and a pocket stereoscope. It is also important that the student begin to understand the relationship between photographs of topographic features and map representations of the same features. Today most topographic maps are made from aerial photos.

Stereoscopic Viewing Of Aerial Photographs

Until the advent of satellite photography, most aerial photographs were taken from airplanes. Today a wide variety of aerial photographs is available, but the most commonly used is still the air photo shot from an airplane with the camera aimed vertically downwards towards the earth's surface. The flight lines and time intervals between shots result in rows of photographs which overlap. This overlap is usually about 60% of the width of any two adjacent photos in one flight line, with about 30% overlap between adjacent lines. Thus every topographic feature is photographed from at least two different viewpoints, which is necessary to achieve a steroscopic or "3-D" image. The length and width dimensions of any topographic feature can be seen on a single photograph. But to see the **relief** (vertical dimension) of the land surface, a stereopair of photos is necessary.

A **stereopair** consists of two adjacent, overlapping photographs from a single flight line. When properly viewed so that each eye sees only one of the photos, the converging lines-of-sight from each eye are perceived by the brain as a single stereoscopic image. Thus the relief of the valleys and ridges may be seen in three dimensions, and, depending on the scale of the photos, the relief may be exaggerated so that hills look steeper and taller than they really are.

The air photos in this exercise are positioned in such a manner as to permit stereoscopic viewing with a small, lens type stereoscope. The three stereopairs on the following pages are best viewed when the center hinge of the stereoscope is positioned over the white line separating a pair of photos. The stereoscope should cross this line at right angles, so that each lens of the stereoscope is over the same topographic feature in each photo. There should be no doubt at all when stereoscopic vision is achieved, for the three dimensional effect is rather dramatic. If stereoscopic vision is not readily achieved, try rotating the stereoscope slightly about its vertical axis. Also, the spacing between the lenses can be adjusted according to the eye-spacing of the viewer.

As the student learns how to view stereo photographs, he should simultaneously be learning to relate topographic features and cultural (man-made) features on the photos to these same features as represented on the accompanying maps.

Answer sheets for this exercise begin on Page A-25.

39

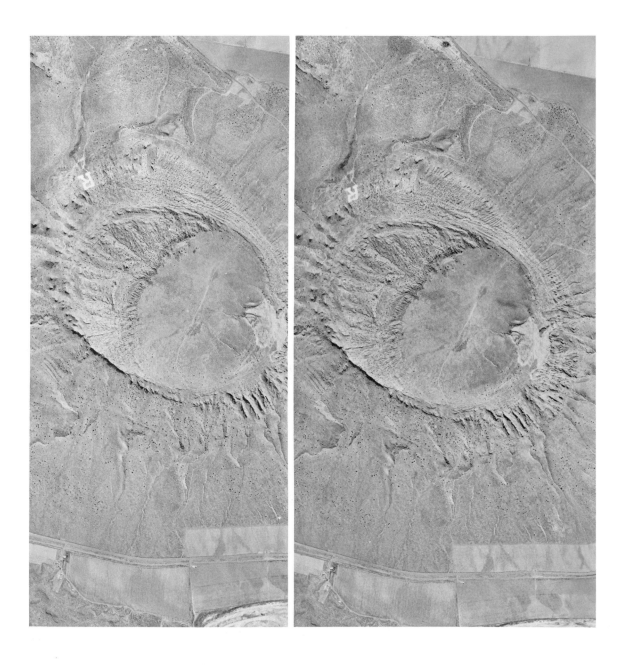

Fig. 9-1. Stereopair of Menan Buttes, Idaho, area

Problems:

1. Menan Buttes are eroded volcanic cinder cones (see discussion of cinder cones in Exercise 19 and refer to Figure 19-1). North Menan Butte, pictured in the stereopair, rises about 800 feet above the plain. Its crater is about a half-mile in diameter and 300-400 feet deep. Notice that the orientation of the photos is not the same as the map orientation. On the map, north is towards the top of the page. Which way is north on the photos? Draw an arrow on one of the photos to indicate the north direction.

2. Compare the steepness of the northeast slope of North Menan Butte at the letter "R" to the western flank of the mountain near the railroad tracks. Which slope is steeper? Now look at North Menan Butte on the topographic map. How are the contour lines (brown lines showing elevation) spaced on the steeper of the two slopes: closer together than on the gentle slope? or further apart than on the gentle slope?

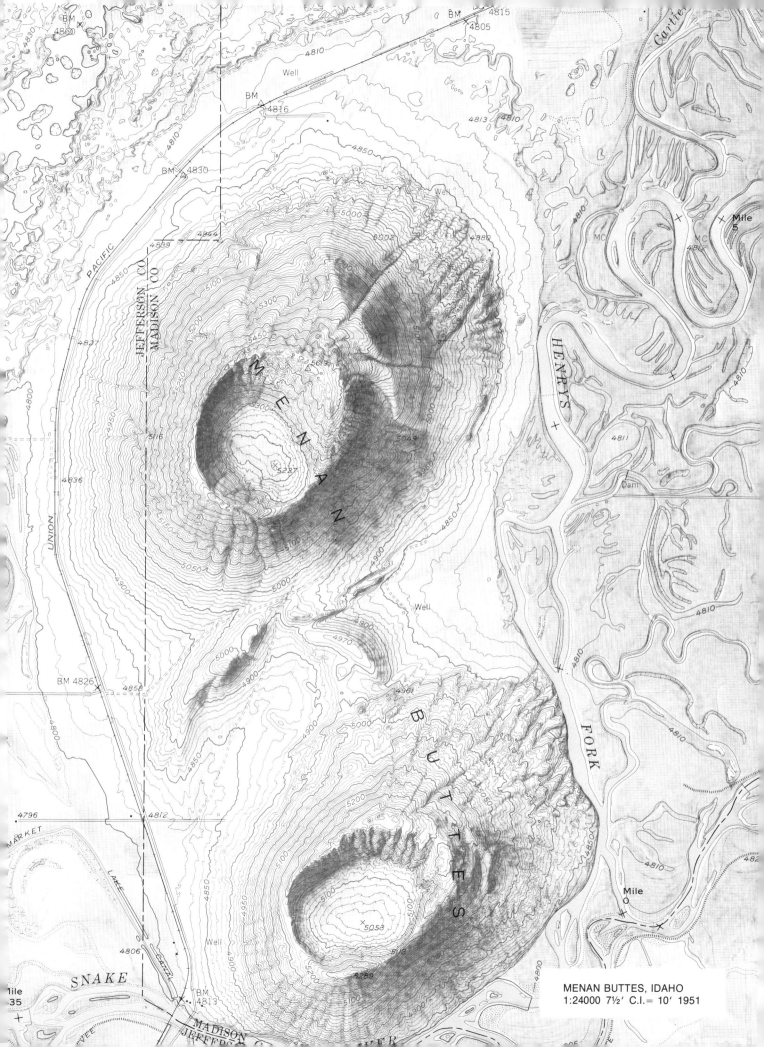

MENAN BUTTES, IDAHO
1:24000 7½' C.I.= 10' 1951

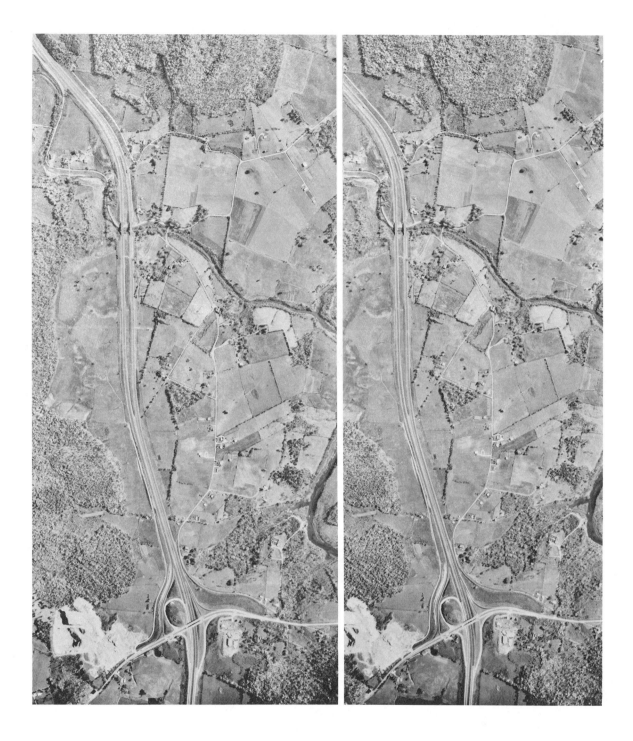

Fig. 9-2. Stereopair of Cookeville East, Tennessee, area

Problems:

1. Locate and mark (by making a neat circle in **pencil)** an example of each of the following features on **both** the stereopair and the Cookeville East map.

    a. bridge over a stream
    b. highway overpass
    c. large rock quarry
    d. Cookeville Filtration Plant for city water
    e. confluence of two streams

COOKEVILLE EAST, TENN.
1:24000 7½' C.I.= 20' 1953

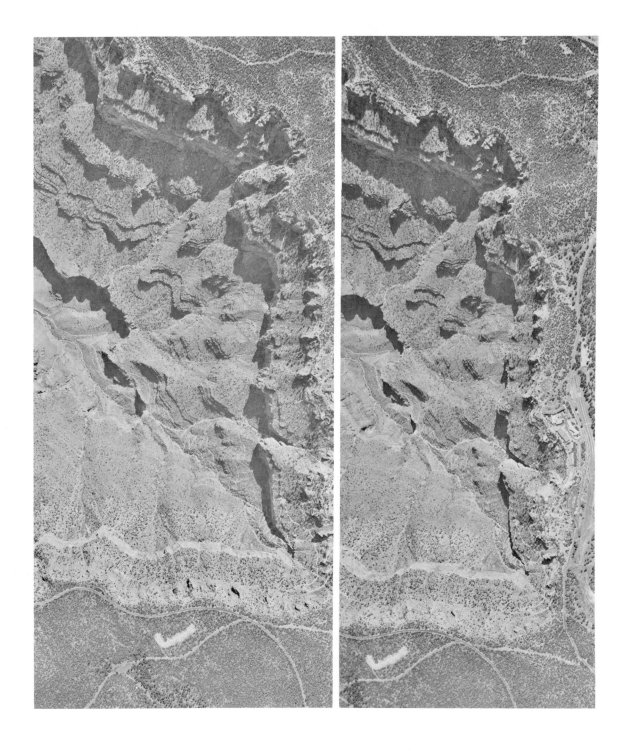

Fig. 9-3. Stereopair of Bright Angel, Arizona, area

Problems:

1. After examining the photos in stereo, write a brief description of the topography pictured in this stereopair. Note the alternating gentle slopes and cliffs due to differential erosion of contrasting lithologies.

2. Which shows the topography in more **detail** (larger scale), the stereopair or the topographic map?

3. In **pencil** draw a box on the map that outlines the exact area shown in the aerial photos.

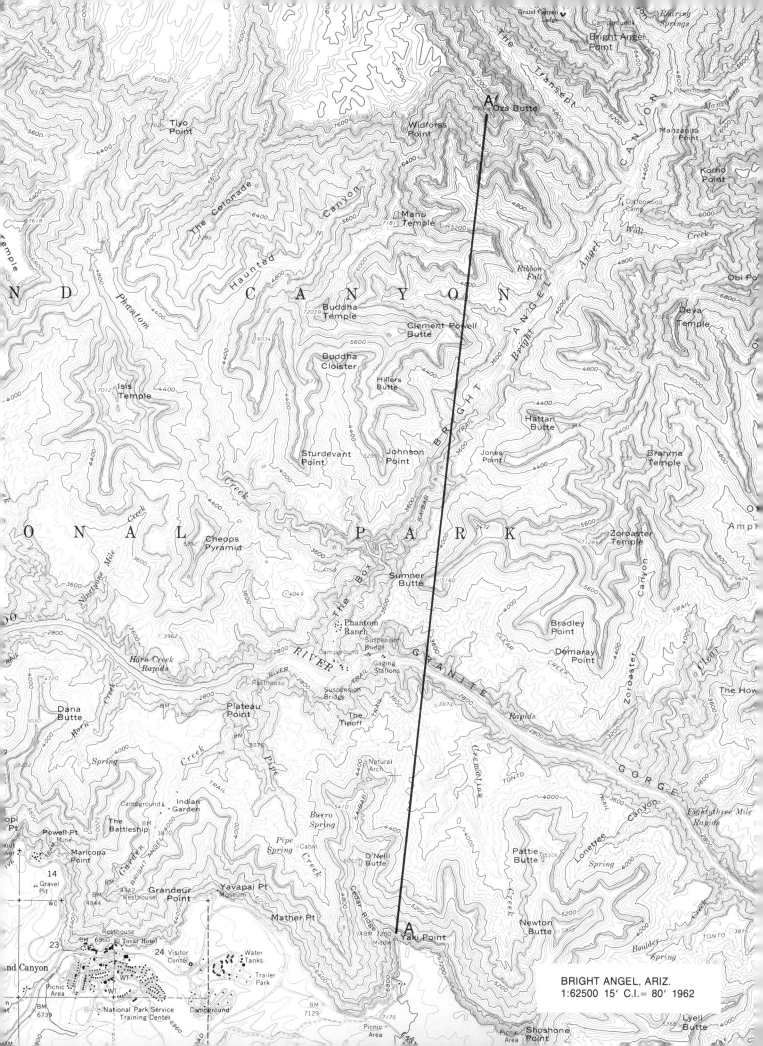

BRIGHT ANGEL, ARIZ.
1:62500 15' C.I.= 80' 1962

Exercise 10: An Introduction to Map Reading

The **topographic map** is one of the most useful tools in the study of the earth's surface and the geological processes affecting it. An understanding of map construction and the ability to read maps correctly is essential for the student of geology. Scales, topographic profiles, map symbols, locational systems, and methods of presenting topographic features are areas that require special competence to facilitate map reading.

The general purpose of most maps is to show the location and distribution of various features on the surface of the earth in their true spatial relations as they would appear if viewed from above, as from an airplane or space capsule directly overhead.

Topographic maps differ from ordinary maps inasmuch as they show **vertical** as well as horizontal spatial relationships. When a land surface is represented by a topographic map, **relief** (differences in elevation) is shown by using various devices such as color, shading, contour lines, raised relief, hachuring, or combinations of these. Therefore topographic maps are two-dimensional representations of three-dimensional surfaces (excepting the case of the raised relief map, which is a three-dimensional model of the real surface). In addition to showing shapes, sizes, and positions of topographic features within the map boundaries, a topographic map shows cultural or man-made objects such as houses, barns, artificial lakes, railroads, pipelines, highways, etc., through use of standard symbols (see USGS Topographic Map Symbols on page 55).

Topographic maps are published by many agencies, including the Department of Defense, Department of Agriculture, the Tennessee Valley Authority, and the U.S. Forest Service, but the major publisher of topographic maps in this country is the Department of Interior through the U.S. Geological Survey whose standard maps are called **quadrangles.** Parallels of latitude form the north and south boundaries and meridians of longitude form the east and west boundary lines of these quadrangles. Presently the most common size is the 7½ minute quadrangle but several different series are still available, including 15 minute, 30 minute, and one degree by two degree series.

Quadrangle maps are named for obvious physical or cultural features on the maps. This practice assists in filing, retrieving, and ordering maps, but requires that no two quandrangles of the same series in the same state can have the same name.

The modern topographic map is made by first surveying, under very exacting conditions, the elevations of a number of carefully selected points that can be identified on **aerial photographs.** Topographic contour lines are then drawn by precise photogrammetric methods using the surveyed points as check points for both horizontal and vertical control. Finally a field check is made to verify the accuracy of the photogrammetric, cultural, drainage, and land grid data.

When effectively used, topographic maps can greatly facilitate the practice of many vocations and avocations. Recreational uses in hiking, orienteering, fishing, hunting, and camping are obvious direct benefits of knowing how to read topographic maps. The general public also benefits indirectly through the use of these maps in geologic mapping, land use planning, agricultural surveys, and highway construction.

Elements Of Topographic Maps

Scale

The scale of a map is an expression of the ratio of distance on the map to the corresponding distance on the ground. The scale chosen for a map or map series will depend upon the size of the area to be mapped, the size of the map sheet desired, and the amount of detail to be shown. Scales are commonly expressed in one of three ways:

1. **Representative fraction (or scale ratio).** Representative fractions are written as 1/12000 (scale ratios are written 1:12000), where the numerator is the map unit and the denominator is the ground equivalent. In other words, one map unit represents 12000 (in this case) of those same units on the ground. Note that the representative fraction (R.F.) has the advantage of being universally applicable. No matter what units are considered, as long as numerator and denominator are expressed in the same units, the R.F. 1/12000 is equally intelligible to Americans thinking in inches, Frenchmen thinking in meters, or Russians thinking in versts.

Beginners in map reading exercises occasionally have difficulty with the representative fraction when discussing "large scale" versus "small scale" maps. The distinction is easily made if the map reader remembers that given two representative fractions, the fraction with the **smaller denominator is the larger scale.** For example, a scale of 1/24000 is larger than a scale of 1/62500.

Large-scale maps show small areas in greater detail, while small-scale maps show large areas in much less detail.

2. **Graphically.** Graphic scales are included on topographic maps in the form of a segmented line (see below) that represents the map to ground relationship.

The line is divided into convenient, appropriate units so that the map user is able to measure distances directly by placing the edge of a piece of paper along the line between any two points on the map. Then, after marking the points on the paper, it is placed along the graphic scale where the distance is read directly.

Another major advantage of the graphic scale is that its relationship to the map remains the same during photographic reduction or enlargement. A fractional scale or a verbal scale becomes meaningless if the map is enlarged or reduced.

3. **Verbally.** The verbal scale is a written or oral statement of the map to ground relationship such as "one inch represents one mile." In this case, one inch on the map represents 63,360 inches on the ground (the number of inches in a mile). Note that the units do not have to be the same although they may be; e.g., "one foot represents twelve thousand feet."

Scale Conversion

Map users are often called upon to convert from one scale type to another. This is easily done after mastering a few fundamental manipulations. A little practice will quickly build confidence and competence. The following examples are typical of those most often necessary:

Fractional scale to verbal scale. Change the representative fraction 1/10000 to the appropriate statement using familiar units such as inches and miles. Set up the simple ratio:

$$\frac{1 \text{ inch on the map}}{10,000 \text{ inches on the ground}} = \frac{\text{map distance}}{\text{ground distance}}$$

First put the ground distance into miles by dividing 10,000 by 63,360, the number of inches in a mile, to get

$$\frac{1 \text{ inch on the map}}{10,000 \text{ inches}/63,360 \text{ inches/mile}} = \frac{1 \text{ inch on the map}}{0.1578 \text{ miles on the ground}}$$

Thus one can say "one inch represents 0.1578 miles" or "6.34 inches represent one mile."

Verbal scale to graphic scale. Making a graphic scale from verbal instructions is usually a simple matter of dividing a line properly. A verbal scale "one inch represents one mile" would be converted by drawing a straight line and dividing it into one-inch segments. The first segment is then subdivided into smaller, more convenient units such as tenths or perhaps fourths.

Occasionally in converting other scales into graphic scales it becomes necessary to divide a line into parts which are not easily taken from a ruler. For example, suppose we are instructed to divide a section of a line into five parts and the line segment to be divided is 4 3/4 inches long. This chore can be done easily with a triangle and a straight edge as follows. Lay out the line A-B so that its length is 4 3/4 inches (See Figure 10-1). Then lay out line A-X at any convenient angle to line A-B. Divide A-X into any five convenient, equal sections (i.e., A-X should be a length easily divisible by 5).

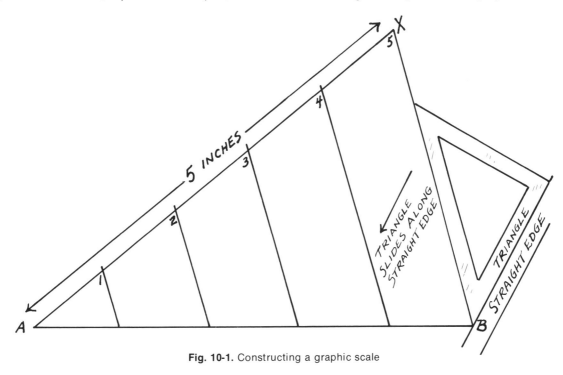

Fig. 10-1. Constructing a graphic scale

Next connect "X" to "B" using the triangle and straight edge as shown. By sliding the triangle along the straight edge and drawing lines from the numbers to line A-B parallel to the line X-B, the line A-B is effectively divided into five equal parts and the graphic scale is complete except for labeling.

Graphic scale to fractional scale. Given the data from the graphic scale that 2.4 inches represents 5 miles (these data will have been measured from the graphic scale using several trials to assure the best possible accuracy), set up the fraction 2.4 inches/5 miles. Convert the miles to inches by multiplying by 63,360 (inches in a mile).

$$2.4 \text{ inches}/5 \text{ miles} \times 63,360 \text{ inches/mile} = \frac{2.4 \text{ inches}}{316,800 \text{ inches}}$$

Reduce this to a simple fraction 1/132000, which is the representative fraction for this map.

Methods of Locating Features on a Map

By custom, the top of the map (when held vertically, or in the normal viewing position) is north, unless specifically marked otherwise. East would then be to the viewer's right, and west to the left. When the top of the map is not north, the north direction should be indicated by an arrow. The direction of any line on a map can be determined by measuring the angle the line makes with any convenient meridian or parallel on the map, or with the north arrow itself.

We are accustomed to the idea that a compass points north. But the compass points toward the earth's north **magnetic pole,** which does not exactly coincide with the north pole of rotation, the **true north** direction. Hence the compass direction at most places on the earth departs from true north. This departure of the compass from true north is known as the *declination of the compass,* or more commonly, **magnetic declination.** For the benefit of compass users in an area covered by a topographic map, it is customary to indicate, graphically, near the bottom of the map, the amount of magnetic declination for the area. The declination may be either east or west, and is shown by a small arrow, sometimes labeled *MN,* which forms an angle with a larger arrow which indicates the true north direction.

Cincinnati, Ohio, is located near the Agonic Line (a line connecting points of no magnetic declination), while the Point Reyes, California, area shows approximately 17½ degrees east declination. The configuration of the isogonic lines (lines of equal declination) shifts slowly with time because the earth's magnetic field is not static or stable. Therefore isogonic charts must be periodically adjusted, and the magnetic declination shown on very old maps may no longer be correct.

Latitude and longitude. Cartographers use a grid system of east-west and north-south trending lines to locate places on the earth's surface. The east-west trending lines are known as **parallels of latitude,** whereas the north-south trending lines are called **meridians of longitude.**

1. **Latitude lines** (parallels) run east-west, parallel to the equator, which is the line of zero degrees latitude. Latitude values are measured in degrees, increasing both north and south away from the equator to maximum values of 90 degrees at the poles. For example, Nashville, Tennessee, lies near the parallel of 36 degrees north latitude (36° N latitude). The distance between any adjacent 1° latitude lines is approximately 69 miles.

2. **Longitude lines** (meridians) run north-south and converge at the poles. As a consequence, the distance between one-degree longitude lines is much greater at the equator (69.17 miles) than at high latitudes near the poles. (For example, at 70° N latitude, the distance between one-degree longitude lines is only 23.73 miles, in contrast to parallels of latitude which are nearly equidistant.) The line of zero degrees longitude passes through Greenwich, England, and is called the **prime meridian.** All longitude is measured in degrees east and west from this line, the maximum longitude value being 180 degrees east or west of the prime meridian. This 180 degree meridian, located on exactly the opposite side of the earth from Greenwich, is the international date line.

Figure 10-2 illustrates the worldwide latitude and longitude grid system. However, the best way to see this relationship is to examine an actual globe.

U.S. Geological Survey topographic maps, and most similar maps issued by other agencies, are printed in rectangular sheets called **quadrangles.** Quadrangle maps are bounded by parallels and meridians, and other lines of latitude and longitude are generally marked at appropriate places along the map borders. A latitude-longitude grid system for the map area is formed this way, making it easy to determine the approximate latitude and longitude of any point shown on the map.

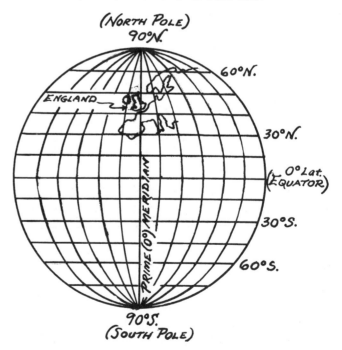

Fig. 10-2. Global latitude and longitude grid system

Each degree of latitude and longitude contains 60 minutes and each minute is subdivided into 60 seconds. A little arithmetic will show that one second of longitude at the equator is equal to approximately 0.0192 miles or 101.45 feet. This obviously permits accurate location by the latitude-longitude method.

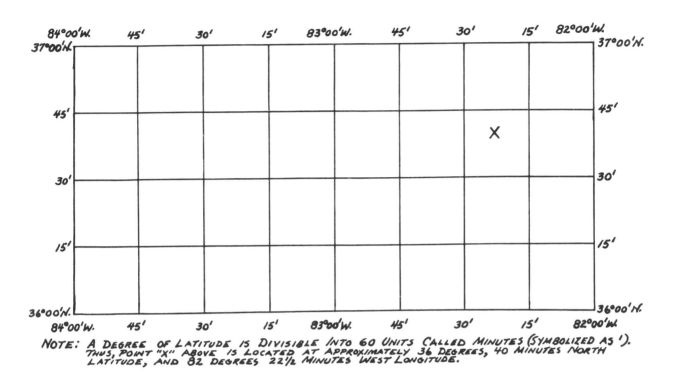

NOTE: A DEGREE OF LATITUDE IS DIVISIBLE INTO 60 UNITS CALLED MINUTES (SYMBOLIZED AS '). THUS, POINT "X" ABOVE IS LOCATED AT APPROXIMATELY 36 DEGREES, 40 MINUTES NORTH LATITUDE, AND 82 DEGREES 22½ MINUTES WEST LONGITUDE.

Fig. 10-3. Latitude-longitude grid taken from Johnson City, Tennessee, 1° x 2° map sheet

Location by rectangles. The parallels and meridians furnish a convenient means of dividing a quadrangle into smaller rectangles for ready, easy reference. For example, the 7½ minute U.S. Geological Survey quadrangles are commonly divided into 9 rectangles, each of which is 2½ minutes on the side. The four corners of rectangle 5 are marked by "crosses" on the map (look for two of these "crosses" on the portion of the Cookeville East Quadrangle, page 43).

For convenience, the rectangles are numbered and, if closer location is desired, each rectangle is again subdivided into 9 smaller rectangles. The process may be repeated as many times as desired. A location near the southwest corner of rectangle 8 would be written 8.7. Point A is in rectangle 9 of rectangle 3, so it would be written 3.9. A second method of designating rectangles is to use letters referring to each subdivision. In this manner, rectangle 1 would be NW, 2 - NC (north-central), 3 - NE, 4 - WC, 5 - C, 6 - EC, 7 - SW, 8 - SC, 9 - SE. Therefore, point A would be located in rectangle SE of rectangle NE, so it would be written NE-SE.

Tier and Range system. It is necessary to have some knowledge of this most common form of land subdivision if the student is to become proficient in locational systems. In 1785 a federal law was enacted for the purpose of subdividing public lands (a cadastral system) into townships and sections; the **townships** were to be six miles square and the sections were to be one mile square. The 36 sections comprising a township were to be numbered from south to north. In 1796 the law was amended so that the sections were to be numbered in east-west rows, alternating east-to-west and west-to-east, starting with section 1 in the northeast corner of the township and ending with section 36 in the southeast. The townships themselves are arranged in east-west rows called **tiers,** and north-south columns called **ranges.** (See Figure 10-4.) This is the system in use today for most of the United States.

The beginning point of a Tier and Range survey is called the *initial point,* which is located where a selected baseline (a convenient parallel of latitude) and a principal meridian (a convenient line of longitude) cross. The exact location of this point is established or determined astronomically. There are 31 initital points in the contiguous 48 states and 3 in Alaska. The Tier and Range system is used in most states west of the Appalachians and most southern states except Texas.

Direction (azimuth or bearing) and distance from a known point. This method of locating points or features involves a statement of directions such as "the sinkhole is located 6½ miles due north of the intersection of Highway 47 and the Cedar Creek Road" or "the camp is 3 miles N 45° W from the old mill site at the mouth of Severn Creek."

State coordinate systems. State coordinate systems have been established for many states. Where footage locations are to be used, a point outside and southwest of the state is chosen. North-south and east-west base lines are then drawn through this selected point. Any point within the state can now be designated as X feet east of and Y feet north of the selected reference point. These footages are shown in the border of some modern large-scale, topographic maps as illustrated below.

If each division equals 1,000,000 feet then point A (below, left) would be designated as 5,000,000 feet east of and 2,000,000 feet north of the reference point.

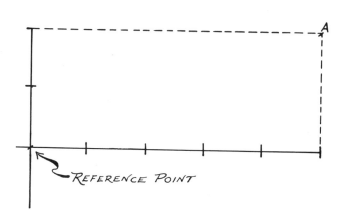
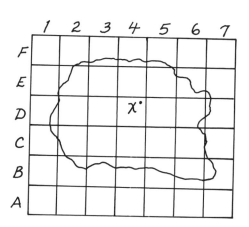

Another state coordinate system involves the use of 5-minute rectangles with number and letter designations. Numbers designate the vertical rows of 5-minute rectangles and letters designate the horizontal rows. Thus, point X (above, right) is designated as D-4. It can be more closely located by measuring the footages from the north line and the east line.

Informal relative reference points. Directions without specific distance units fall into this category; e.g., the mulberry tree is near Morgadore on Big Cedar Creek or the leaking pond is behind the barn across from the old school building on Big Twin Creek.

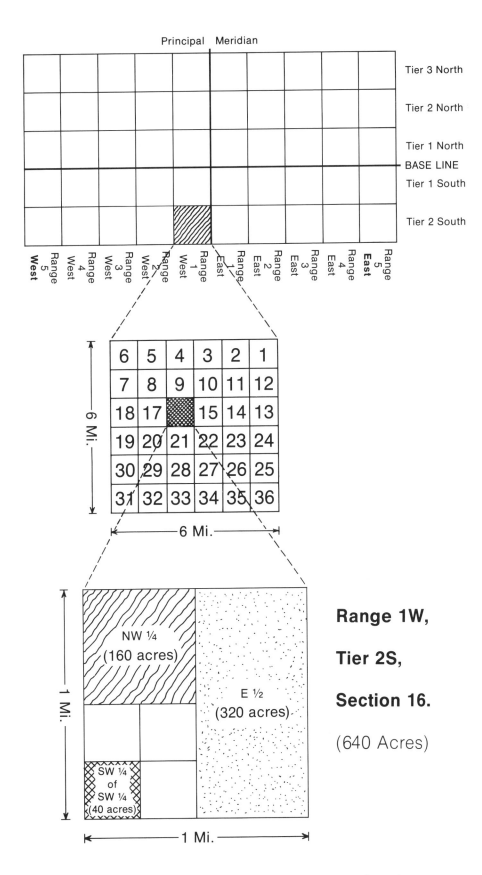

Fig. 10-4. Public land survey of the U.S.A. (Tier and Range System)

Symbols and Colors On Topographic Maps

The use of a reasonably standardized, easily understood set of map symbols in conjunction with a few standard colors makes it possible to present vast quantities of data on topographic maps without objectionable clutter or feature overlap. A standard guide to topographic map symbols and color keying used by the U.S. Geological Survey is included in this exercise. The student should become familiar with the contents of this guide.

Briefly, map features are differentiated as follows: **black** is used to show buildings, some roads, high voltage power lines, cemeteries, political boundaries; **brown** is used to show landforms, contour lines, embankments, earth-fill dams, mine dumps and other man-made relief features; **blue** is reserved for water features including ponds, lakes, streams, oceans, canals, etc., and is sometimes used for glaciers, although white with a bluish tint is also used for glaciers (water features are also identified by use of a special kind of type called *hydrologic type* when the feature has a specific name); **green** is used for woodlands and forests with special patterns for orchards, vineyards, tree farms, and scrub growth; **red** is used for major highways, public land divisions, fence lines, and in some cases has been utilized to emphasize the numbers in a tier and range system shown on the map; **pink** or **pale red** designates urban areas within which major landmarks are identified by name or perhaps by shape alone such as a race track; finally, **purple** indicates photorevision of an older map, where the revision is overprinted in purple to show major changes since the map was published. The photo revision is done in the office and is not field checked for accuracy.

Turn to Page A-27 for questions on material covered in this exercise.

AN AZIMUTH COMPASS SUCH AS THAT USED BY LEWIS AND CLARK.

POCKET SEXTANT OF LATE 1700'S.

50-FOOT LINEN MEASURING TAPE USED BY SURVEYORS OF THE 1800'S.

PRISMATIC COMPASS OF LATE 1800'S.

SOME MAPPING INSTRUMENTS
OF
THE 18TH AND 19TH CENTURIES

TOPOGRAPHIC MAP SYMBOLS

VARIATIONS WILL BE FOUND ON OLDER MAPS

Primary highway, hard surface . . . .

Secondary highway, hard surface . . . .

Light-duty road, hard or improved surface . . . .

Unimproved road . . . .

Road under construction, alinement known . . . .

Proposed road . . . .

Dual highway, dividing strip 25 feet or less . . . .

Dual highway, dividing strip exceeding 25 feet . . . .

Trail . . . .

Railroad: single track and multiple track . . . .

Railroads in juxtaposition . . . .

Narrow gage: single track and multiple track . . . .

Railroad in street and carline . . . .

Bridge: road and railroad . . . .

Drawbridge: road and railroad . . . .

Footbridge . . . .

Tunnel: road and railroad . . . .

Overpass and underpass . . . .

Small masonry or concrete dam . . . .

Dam with lock . . . .

Dam with road . . . .

Canal with lock . . . .

Buildings (dwelling, place of employment, etc.) . . . .

School, church, and cemetery . . . . Cem

Buildings (barn, warehouse, etc.) . . . .

Power transmission line with located metal tower . . . .

Telephone line, pipeline, etc. (labeled as to type) . . . .

Wells other than water (labeled as to type) . . . . o Oil . . . . . o Gas

Tanks: oil, water, etc. (labeled only if water) . . . . ● ● ● ⊘ Water

Located or landmark object; windmill . . . . o

Open pit, mine, or quarry; prospect X

Shaft and tunnel entrance . . . . ▪ Y

Horizontal and vertical control station:

    Tablet, spirit level elevation . . . . BM △ 5653

    Other recoverable mark, spirit level elevation . . . . △ 5455

Horizontal control station: tablet, vertical angle elevation VABM △ 9519

    Any recoverable mark, vertical angle or checked elevation △ 3775

Vertical control station: tablet, spirit level elevation . . . . . BM × 957

    Other recoverable mark, spirit level elevation . . . . × 954

Spot elevation . . . . × 7369   × 7369

Water elevation 670

Boundaries: National . . . .

    State . . . .

    County, parish, municipio . . . .

    Civil township, precinct, town, barrio . . . .

    Incorporated city, village, town, hamlet . . . .

    Reservation, National or State . . . .

    Small park, cemetery, airport, etc. . . . .

    Land grant . . . .

Township or range line, United States land survey . . . .

Township or range line, approximate location . . . .

Section line, United States land survey . . . .

Section line, approximate location . . . .

Township line, not United States land survey . . . .

Section line, not United States land survey . . . .

Found corner: section and closing . . . .

Boundary monument: land grant and other . . . . □ □

Fence or field line . . . .

Index contour		Intermediate contour . .	
Supplementary contour		Depression contours . .	
Fill		Cut	
Levee		Levee with road	
Mine dump		Wash	
Tailings		Tailings pond	
Shifting sand or dunes		Intricate surface	
Sand area		Gravel beach	

Perennial streams		Intermittent streams . .	
Elevated aqueduct		Aqueduct tunnel	
Water well and spring . . .		Glacier	
Small rapids		Small falls	
Large rapids		Large falls	
Intermittent lake		Dry lake bed	
Foreshore flat		Rock or coral reef	
Sounding, depth curve . .		Piling or dolphin	o
Exposed wreck		Sunken wreck	⊬
Rock, bare or awash; dangerous to navigation			✳ ⊛

Marsh (swamp)		Submerged marsh	
Wooded marsh		Mangrove	
Woods or brushwood . . .		Orchard	
Vineyard		Scrub	
Land subject to controlled inundation		Urban area	

UNITED STATES
DEPARTMENT OF THE INTERIOR
GEOLOGICAL SURVEY

TOPOGRAPHIC
MAP INFORMATION AND SYMBOLS
SEPTEMBER 1972

QUADRANGLE MAPS AND SERIES

Quadrangle maps cover four-sided areas bounded by parallels of latitude and meridians of longitude. Quadrangle size is given in minutes or degrees. The usual dimensions of quadrangles are: 7.5 by 7.5 minutes, 15 by 15 minutes, and 1 degree by 2 or 3 degrees.

Map series are groups of maps that conform to established specifications for size, scale, content, and other elements.

MAP SCALE DEPENDS ON QUADRANGLE SIZE

Map scale is the relationship between distance on a map and the corresponding distance on the ground.

Map scale is expressed as a numerical ratio or shown graphically by bar scales marked in feet, miles, and kilometers.

NATIONAL TOPOGRAPHIC MAPS

Series	Scale	1 inch represents	Standard quadrangle size (latitude-longitude)	Quadrangle area (square miles)	Price
7½-minute	1:24,000	2,000 feet	7½×7½ min.	49 to 70	$0.75
Puerto Rico 7½-minute	1:20,000	about 1,667 feet	7½×7½ min.	71	.75
15-minute	1:62,500	nearly 1 mile	15×15 min.	197 to 282	.75
Alaska 1:63,360	1:63,360	1 mile	15×20 to 36 min.	207 to 281	.75
U. S. 1:250,000	1:250,000	nearly 4 miles	1° × 2° or 3°	4,580 to 8,669	1.00
U. S. 1:1,000,000	1:1,000,000	nearly 16 miles	4° × 6°	73,734 to 102,759	1.00
Antarctica 1:250,000	1:250,000	nearly 4 miles	1° × 3° to 15°	4,089 to 8,336	.75
Antarctica 1:500,000	1:500,000	nearly 8 miles	2° × 7½°	28,174 to 30,462	.75

CONTOUR LINES SHOW LAND SHAPES AND ELEVATION

The shape of the land, portrayed by contours, is the distinctive characteristic of topographic maps.

Contours are imaginary lines following the ground surface at a constant elevation above or below sea level.

Contour interval is the elevation difference represented by adjacent contour lines on maps.

Contour intervals depend on ground slope and map scale; they vary from 5 to 1,000 feet. Small contour intervals are used for flat areas; larger intervals are used for mountainous terrain.

Supplementary dotted contours, at less than the regular interval, are used in selected flat areas.

Index contours are heavier than others and most have elevation figures.

Relief shading, an overprint giving a three-dimensional impression, is used on selected maps.

Orthophotomaps, which depict terrain and other map features by color-enhanced photographic images, are available for selected areas.

COLORS DISTINGUISH KINDS OF MAP FEATURES

Black is used for manmade or cultural features, such as roads, buildings, names, and boundaries.

Blue is used for water or hydrographic features, such as lakes, rivers, canals, glaciers, and swamps.

Brown is used for relief or hypsographic features—land shapes portrayed by contour lines.

Green is used for woodland cover, with patterns to show scrub, vineyards, or orchards.

Red emphasizes important roads and is used to show public land subdivision lines, land grants, and fence and field lines.

Red tint indicates urban areas, in which only landmark buildings are shown.

Purple is used to show office revision from aerial photographs. The changes are not field checked.

INDEXES SHOW PUBLISHED TOPOGRAPHIC MAPS

Indexes for each State, Puerto Rico and the Virgin Islands of the United States, Guam, American Samoa, and Antarctica show available published maps. Index maps show quadrangle location, name, and survey date. Listed also are special maps and sheets, with prices, map dealers, Federal distribution centers, and map reference libraries, and instructions for ordering maps. Indexes and a booklet describing topographic maps are available free on request.

HOW MAPS CAN BE OBTAINED

Mail orders for maps of areas east of the Mississippi River, including Puerto Rico, the Virgin Islands of the United States, and Antarctica should be ordered from the U. S. Geological Survey Distribution Section, 1200 South Eads Street, Arlington, Virginia 22202. Maps of areas west of the Mississippi River, including Alaska, Hawaii, Louisiana, Minnesota, American Samoa, and Guam should be ordered from the Distribution Section, Federal Center, Denver, Colorado 80225. A single order combining both eastern and western maps may be placed with either office. Residents of Alaska may order Alaska maps or an index for Alaska from the Distribution Section, 310 First Avenue, Fairbanks, Alaska 99701. Order by map name, State, and series. Maps without woodland overprint are available on request. On an order amounting to $300 or more at the list price, a 30-percent discount is allowed. No other discount is applicable. Prepayment is required and must accompany each order. Payment may be made by money order or check payable to the U. S. Geological Survey, or cash (the exact amount) at sender's risk. Your ZIP code is required.

Sales counters are maintained in the following U. S. Geological Survey offices, where maps of the area may be purchased in person: 1200 South Eads Street, Arlington, Va.; Room 1028, General Services Administration Building, 18th & F Streets N W., Washington, D. C.; 1109 North Highland Street, Arlington, Va.; 900 Pine Street, Rolla, Mo.; 345 Middlefield Road, Menlo Park, Calif.; 7638 Federal Building, 300 North Los Angeles Street, Los Angeles, Calif.; 504 Custom House, 555 Battery Street, San Francisco, Calif.; Building 41, Federal Center, Denver, Colo.; 1012 Federal Building, 1961 Stout Street, Denver, Colo.; Room 1-C 45, 1100 Commerce Street, Dallas, Texas; 8102 Federal Building, 125 South State Street, Salt Lake City, Utah; 678 U. S. Court House, West 920 Riverside Avenue, Spokane, Wash.; 108 Skyline Building, 508 Second Avenue, Anchorage, Alaska; 441 Federal Building, 709 West Ninth Street, Juneau, Alaska; and 310 First Avenue, Fairbanks, Alaska.

Commercial dealers sell U. S. Geological Survey maps at their own prices. Names and addresses of dealers are listed in each State index.

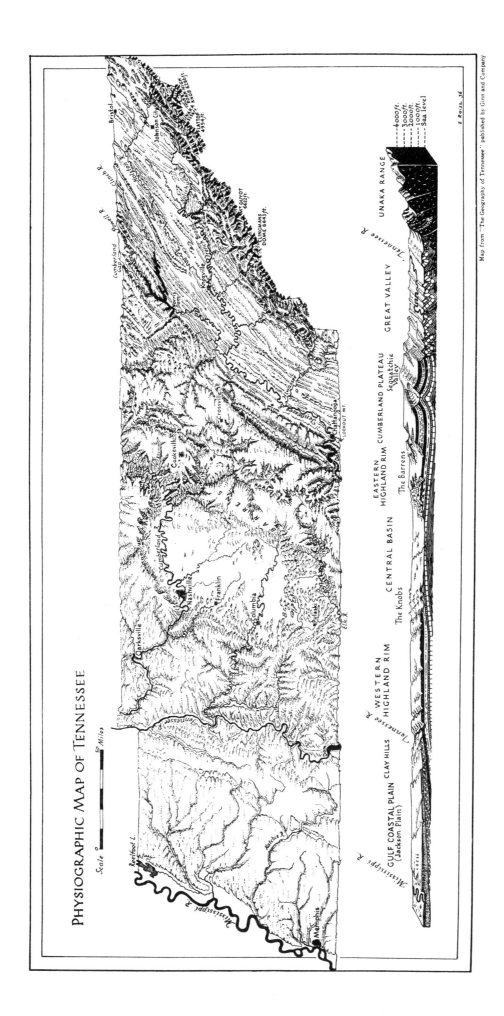

PHYSIOGRAPHIC MAP OF TENNESSEE

Scale 0 — 50 Miles

Map from "The Geography of Tennessee" published by Ginn and Company

E Raisz. 36

4000 ft.
3000 ft.
2000 ft.
1000 ft.
Sea level

UNAKA RANGE

GREAT VALLEY

CUMBERLAND PLATEAU

EASTERN HIGHLAND RIM

CENTRAL BASIN

WESTERN HIGHLAND RIM

GULF COASTAL PLAIN CLAY HILLS
(Jackson Plain)

Sequatchie Valley

The Barrens

The Knobs

Mississippi R.
Tennessee R.

Bristol
Johnson City
Memphis
Clarksville
Nashville
Franklin
Columbia
Pulaski
Cookeville
Crossville
Chattanooga
Lookout Mt.
Knoxville
Norris Dam

LINGMANS DOME 6643 ft.
GUYOT 6621 ft.
CATTOP KNOB 5934 ft.

Reelfoot L.
Hatchie R.
Elk R.
Cumberland R.
Tennessee R.
Clinch R.
Powell R.
Holston R.
French Broad R.

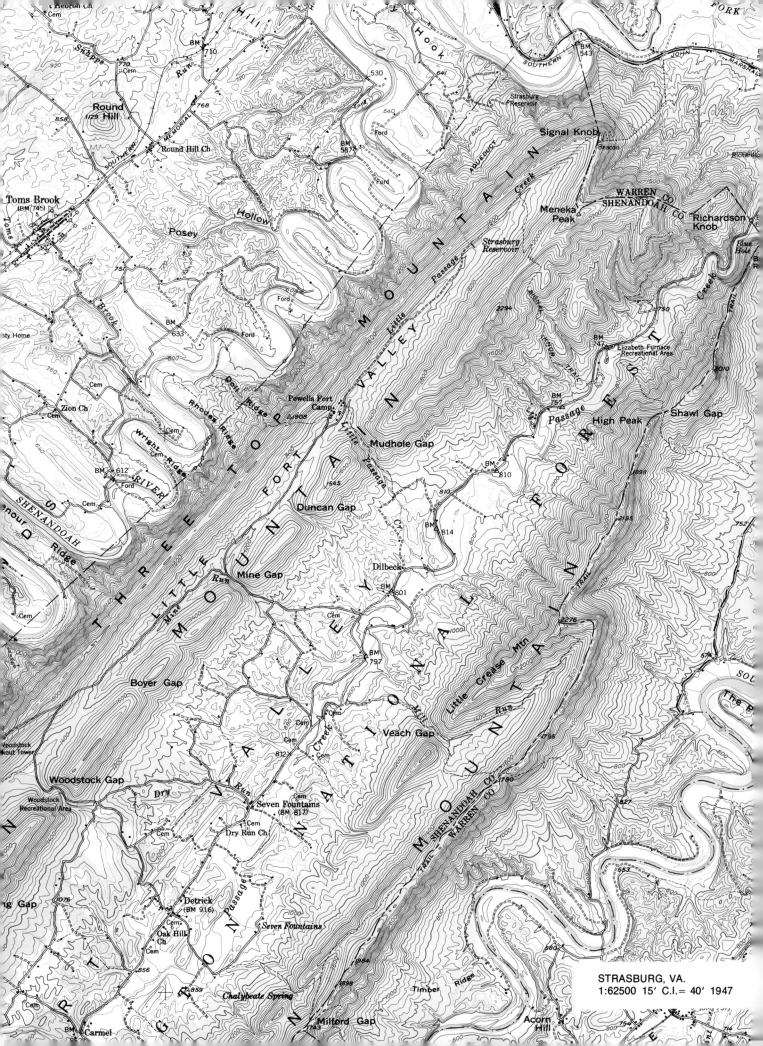

STRASBURG, VA.
1:62500   15'   C.I.= 40'   1947

Exercise 11: Methods of Representing Topography

This exercise is concerned with the common methods of representing topography and the construction of topographic maps.

Altitude

The altitude of a specific point is its elevation above some well-recognized datum plane, usually sea level.

Relief

The local difference in altitude is the **relief.** For example, if valleys of a region have an altitude of about 500 feet and the hilltops have an altitude of about 800 feet, the relief is approximately 300 feet. Maximum relief is obtained for any one map or area by subtracting the lowest elevation from the highest elevation.

Slope or Gradient

The inclination of the surface of the land, a road, or a stream channel is described as the **slope** or **gradient,** and may be expressed in three ways: (1) feet per mile (or meters per kilometer in the metric system); (2) in percent; and (3) in degrees.

Gradient is most commonly expressed in feet per mile, and may be calculated using the following equation:

$$\text{Gradient} = \frac{\text{Elevation change in feet}}{\text{Horizontal distance in miles}}$$

Note: In computing stream gradient, the horizontal distance must be measured along the stream channel.

Topography

It can readily be seen that some means of representing hills, valleys, plains, and other topographic features on a flat sheet of paper is necessary. Following are a number of methods used on modern maps.

Plastic Relief Models. Topography may be represented by a relief model which is a small-scale replica of the original land surface. The model is probably the easiest to comprehend and is very suitable for exhibiting in classrooms and museums, but it is usually more inaccurate than most other methods and the vertical exaggeration (see Exercise 12) of the model distorts the true relationship in order to show relief. Furthermore, a relief model is not convenient for use in many cases because it is not a flat sheet of paper which can be folded or rolled for easy carrying or storing.

Hachures. Hachures are short lines drawn on a map parallel to the slope of the land, from higher areas to lower (see the physiographic map of Tennessee, page 57). Steep slopes are represented by heavier lines, more closely spaced and longer than those used for gentler slopes. Hachures are advantageous for representing general forms where exact elevations are not known or not necessary. Hachure maps are not very accurate as exact elevations are not shown, except for some spot elevations on some maps.

Color Layering. For small-scale maps where it is desired that the topography be visible from a distance (e.g., for wall maps for classroom use), elevation ranges are indicated by different shades of a color or colors. For instance, all areas lying between sea level and 500 feet above sea level could be colored dark green; from 500 to 1000 feet, light green; 1000 to 1500 feet, light tan; from 1500 to 2000 feet, dark tan; etc. The exact elevation of points within color bands is not known. Any point exactly on a color boundary would have a known value, as the color boundary is in reality a contour line.

Shading. Topography may be portrayed by shading in much the same manner as that utilized by artists to give the illusion of the third dimension. The shading is usually on the southeast side of topographic highs, as if the light source were to the northwest. The darkened areas would be in the shadow of the higher elevations. Shading is highly graphic and easy to understand, and when coupled with contours which give exact elevations, the two make a most useful, effective, and easily understood map (refer to the Menan Buttes Quadrangle, Idaho, page 41).

Contours. A contour is a line on a map which connects points of equal elevation. The contour line represents a certain selected elevation, and **all** points on the contour line must have that elevation. The shoreline of a lake would be an example of such a line. If we could lower the level of the lake by 10 feet, then a second contour line could be drawn at the new shoreline, and the **contour interval** (C.I.) would be ten feet. This could be repeated until a whole series of contour lines had been drawn, and we would then have a contour map showing the topography of the land formerly covered by our partially drained lake.

It is customary to choose a definite interval between elevations to be represented by the contours. In the United States, contour intervals of 2, 5, 10, 20, 40, 50, 100, 250, and 500 feet are commonly used, according to the relief of the area and the scale of the map.

The elevation of any point on a contour line is, of course, the elevation represented by that contour. Any point between contour lines is estimated on the assumption that the rate of slope is uniform. A point halfway between the 20 and 40 foot contours would be estimated as 30 feet. A point one-fourth of the way would be 25 or 35 feet (depending on whether one was going up or down).

Bathymetric contours represent water depth below lake level or sea level and are commonly expressed in feet or in fathoms (one fathom equals 6 feet).

Fig. 11-1. Relationship between landforms and topographic contour map

Rules of Contouring. (Adapted from Brownlow and Reinhard, 1968, *Laboratory Manual, Geology for Engineers:* Wm. C. Brown Co., Publishers, pp. 25-26).

1. A given contour line can represent one, and only one, elevation.
2. Every contour line closes (meets itself enclosing a somewhat circular area) somewhere, although it may not be shown on the map.
3. Contour lines do not cross. (Contour lines under an overhang appear to cross other contour lines but the hidden lines are dashed so they may be distinguished, or they may appear to merge temporarily with higher lines.)
4. Contour lines never merge or divide, except as noted above.
5. Adjacent contour lines of the same elevation do not, under most natural circumstances, exactly parallel one another.
6. Relatively speaking, on the same map, close spacing of adjacent contours indicates a steep slope, and wide spacing of adjacent contours indicates a gentle slope.
7. The presence of contour lines always indicates a slope.
8. When contours cross a stream valley, they bend upstream forming a V-shape, the point of the V pointing upstream or upvalley. (See Problem 8 in answer sheets for this exercise.)
9. Contours always cross streams or rivers at right angles.
10. Depressions lacking surface outlets are designated by hachures perpendicular to the contour line and pointing in the downslope direction.
11. When the land surface goes uphill and then into a depression, the first hachured contour line has the same value as the last normal contour line crossed.
12. When the land surface goes downhill and into a depression, the first hachured contour line is one contour interval less than the last normal contour line crossed.
13. Index contour lines (every fifth or sometimes every fourth contour line) are printed darker, and are numbered with the elevations.
14. Normally, only one contour interval is used on a map, but more than one may be used when two or more highly contrasting terrains are shown by the same map.
15. When one closed normal contour line is surrounded by another, the inner contour indicates a higher elevation than the other.
16. When one closed hachured contour line is surrounded by another, the inner hachured contour indicates a lower elevation than the other.
17. Blue contour lines along lake shores or other bodies of water illustrate water depth, and are called *bathymetric* lines. Do not mistake decorative form lines paralleling shorelines on older maps for bathymetric lines.

Constructing A Contour Map From Data Points

Before the development of modern photogrammetric methods and the extensive use of aerial photographs, topographic maps (contour maps) were constructed in the field where point elevations were surveyed with field equipment and the contour lines were then sketched in the field. While this method is slow and cumbersome and no longer used, except for special situations or needs or when photographic control is not available, knowledge of the principles involved will greatly assist in understanding topographic maps.

To construct a topographic map from selected data points, it is necessary to follow the rules of contouring listed previously and to keep in mind the following five factors:

1. All elevation points on the map must be honored. No one point is more or less valid than another.

2. Fixed contour intervals make the map more useful and more easily interpreted.

3. Start at the lower elevations and work toward the major streams if any exist in your map area.

4. Interpolate between two known elevation points to establish the position of a contour line when the contour value falls between the value of the two points; e.g., if the points have elevations of 45 and 55 feet respectively, then assume a uniform slope, and place the 50-foot contour halfway between the two points. Note that **no contour line can be drawn between two adjacent points unless one of the points has a value more than that of the contour line being drawn and the other has a value less than that of the contour line being drawn.** For example, on a map with a 10-foot interval, no contour can be drawn between two adjacent points with respective values of 41 and 49 feet. However, a contour line **must** be drawn between two points with values of 48 and 56 feet.

5. Make no assumptions on map data. Contour the data exactly as shown on the map (industry standards permit one contour above and one contour below the last known point of control).

Turn to Page A-31 for questions on material covered in this exercise.

Exercise 12: Topographic Profiles

The contours on a topographic map represent, as discussed in Exercises 10 and 11, landforms as seen from above or a "bird's-eye" view. A more familiar view for most of us non-fliers is the **profile** or side view of hills, valleys, etc. A **topographic profile** shows the intersection of the land surface with a vertical plane. A profile can be constructed along any designated line on a topographic map and will present an accurate graphic representation of the landforms if the horizontal scale equals the vertical scale. However, in order to show detail, topographic profiles are commonly drawn with the vertical scale several times greater than the horizontal scale. This difference in scales is termed **vertical exaggeration** and is arrived at by dividing the vertical scale into the horizontal scale (see explanation under **Scale).** Figure 12-1 shows a typical topographic profile. It is properly constructed, oriented, and shows the *ABC's* of profile construction. **The student should refer to it while studying the following descriptive material.**

Topographic profiles should be drawn with as little vertical exaggeration as possible while still showing the necessary features for proper interpretation of landforms. The reason for this is that large vertical exaggerations tend to make "mountains out of molehills" and a distorted representation of the true landforms is the result.

A well-prepared topographic profile should include the following:

Title. The title should be placed either above or below the profile in an appropriate position so as not to hide or clutter other data; it should show something about what landform or feature is being profiled, where it is to be found, and if convenient the orientation of the profile. Example: "East - West Topographic Profile of the Front Range - Hogbacks Area North of Fort Collins, Colorado."

Scale. Horizontal scale, vertical scale, and vertical exaggeration should be listed together somewhere below or to the side of the profile. The horizontal scale is the scale of the map and cannot be varied without some sort of enlarger-reducer device such as a photo-enlarger. The vertical scale may be varied to suit the purpose of the profiler but must be recorded, along with how much exaggeration has been used. In order to calculate vertical exaggeration (V.E.) both horizontal and vertical scales must be in the same units. It is usually best if both scales are in the representative fraction form.

Example: Horizontal scale = 1/62500 (a representative fraction)
Vertical scale one inch represents 1000 feet (a verbal scale)

To calculate V.E., first convert the vertical scale into a representative fraction — "One inch represents 1000 feet" is the same as saying "One inch represents 12000 inches"; i.e., 1″ = 12000″ which is the same thing as 1/12000, a representative fraction. The units will now be the same for both horizontal and vertical scales, so that vertical exaggeration can be calculated as follows:

$$\text{V.E.} = \frac{\text{horizontal scale}}{\text{vertical scale}} = \frac{62500}{12000} = 5.2\text{X}$$

This means that, in this case, the vertical aspects of the profile are magnified 5.2 times.

Orientation Labels. The compass directions at each end of the profile line should be labeled. By standard convention topographic profiles are **oriented as if viewed from the south;** therefore, on an east-west profile, west will be to the viewer's left and east to the right. The viewer will, of course, be looking north. Continuing with this convention, a NW-SE profile will have the NW end to the left and the SE end to the right. Note that the convention breaks down if the profile is oriented due N - S in which case it is probably best to put N on the left and S on the right. Keep in mind that this guide to profile orientation can be disregarded anytime there is a good reason for doing so, as long as the viewer is alerted to the change by proper orientation labels.

Another word about directions on profiles: profiles differ from maps in that they have only two compass directions (whereas maps represent all the compass directions). Furthermore, profiles have a vertical direction which is lacking in maps. *Straight up* on a profile really represents straight up in the vertical sense, but *straight up* on a map usually means north, although the words *straight up* are not truly applicable to a map.

Graphic Vertical Scale. A graphic vertical scale should be constructed at **both ends** of the topographic profile line; this not only makes it easier to actually construct the profile, but it makes the profile more meaningful to the viewer. These graphic scales should be labeled in either feet or meters above sea level. (Using an English scale map means that most of the vertical scales must be expressed in feet.) Sea level (zero elevation) may or may not appear on the graphic vertical scale. As long as the lower value on the vertical is less than the lowest elevation on the profile, one need not project the base of the graphic vertical scale to zero. The graphic scales should be an integral part of the box or frame which outlines the topographic profile and makes it appear neater and more pleasing to the eye.

Prominent Landmarks. Prominent features such as mountain peaks, valleys, rivers, towns, lakes, etc., should be labeled on the profile for the guidance of the viewer in relating the profile to the map. Care should be taken that the profile outline does not become cluttered with superfluous names.

Suggestions to Make Drawing Topographic Profiles Easier

1. Use graph paper (English scale paper if map scale is in English units; metric scale paper if metric scale units appear on the map).

2. Fold over the edge of the graph paper so that the graph lines reach the edge of the paper and then place the graph paper directly on the map (see Figure 12-1B). Data may thus be transferred directly from the map to the graph paper. This will improve accuracy and make an intermediate step unnecessary.

3. Mark the **end points** of the profile on the graph paper first; then if the paper happens to slip or shift you can easily relocate it properly. (If your paper slips and you do not reposition it correctly, the profile will be distorted and incorrect.)

4. Be sure you record sufficient numbers of points to give your profile accuracy. This generally means include all index contour lines as data points plus some of the intermediate contour lines. These intermediate lines must be used to determine hillcrests and valley bottoms accurately, and may also be used when topography is gentle or relief is so low that there are not enough index contour lines to give profile accuracy or definition.

5. Devise some personal, convenient system of marking your contour line data points so you will know what each point indicates — know, for example, when the elevation direction is changing from upslope to downslope or what the elevation values of the data points are, etc.

6. Work neatly and carefully, erase cleanly, and try to visualize the landforms as you construct the topographic profile.

Turn to Page A-37 for questions on material covered in this exercise.

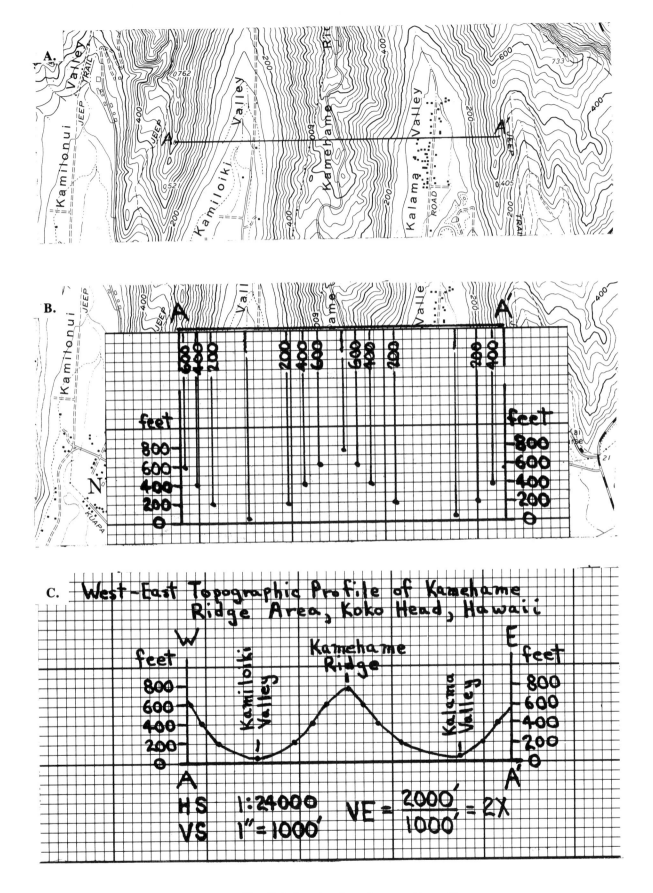

Fig. 12-1. Constructing a topographic profile

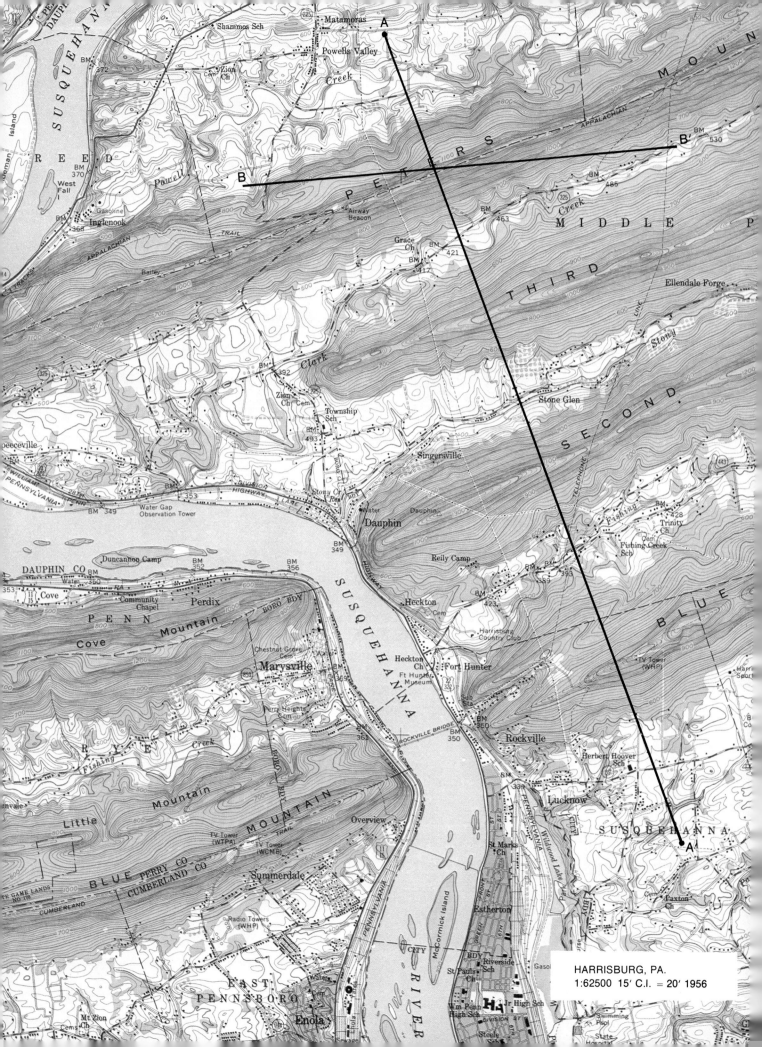

HARRISBURG, PA.
1:62500 15' C.I. = 20' 1956

Exercise 13: Streams, Stream Systems and Topography

Of all the geologic agents which shape the surface of the land, none is more obvious or ubiquitous than running water. With few exceptions (such as areas covered by continental ice sheets or the rare areas entirely dominated by windblown sand), running water has been the dominant agent which has shaped the land surface. Most land areas are, in fact, sloping surfaces which form part of a stream drainage. This is true even in very arid areas; the common notion that deserts are typified by sand dunes is erroneous, for even desert areas are characterized by stream valleys and interstream slopes because water is the dominant agent of erosion.

The climatological and geological factors which determine how a stream develops, how it changes and grows through time, are many and varied. Obviously the amount of rainfall affects the character of a stream. Not so obviously, the development of a stream or drainage system will be affected by the geology of the area: the bedrock units into which the stream valley is cut (some rocks are highly resistant to erosion, others are not), the geologic structures underlying the area (whether the rock units are layered or non-layered, flat-lying, tilted, or folded, fractured or non-fractured), and many other geologic factors.

Because a stream is a physical system which responds to natural physical laws (e.g., water flows downhill because of gravity), the evolution of a stream system should be predictable. But the total number of factors involved is so high that a stream system is not easy to analyze.

Theoretically, a stream may achieve a state of **equilibrium,** wherein the **gradient, velocity, volume, sediment load,** and **general channel characteristics** are precisely adjusted so that no further changes in the system occur; i.e., the stream is neither eroding nor depositing, but simply transporting the sediments already in the stream channel. In reality streams do not achieve such a permanent equilibrium, because any change in one of the controlling factors necessitates compensatory changes in the others. For example, heavy rains upstream from a given point will increase the volume of the stream at that point. This will cause the velocity to increase, so the stream now has both a greater **capacity** and **competency.** It will therefore begin to actively modify its gradient and channel by erosion, in order to pick up more and larger sediments to satisfy the increased capacity and competency. After the heavy rains have ceased, the stream will have to readjust again, but some changes will be irrevocable: the channel characteristics will never be quite the same again. At best then, a stream may achieve a state of **dynamic equilibrium,** in which the controlling factors are in a state of very delicate near-balance, but the balance is constantly shifting, and the stream system constantly evolving.

Attempts have been made to classify the evolutionary stages of stream development. Any such generalization involves a partial departure from reality, but even so, valuable concepts may be illustrated by generalizations. The block diagrams in Figure 13-1 illustrate the hypothetical development of a stream system and the concurrent evolution of the land surface due to stream erosion and deposition. In studying this model, several things should be remembered: (1) the model is a generalization; (2) the terms *youthful, mature,* and *old age* are used with no fixed time span implied (a youthful stream in one region may be millions of years older than a mature stream in a region of different climatological/geological characteristics); (3) a single stream may be in the youthful stage of development near its head-waters, in the mature stage further downstream, and in the old age stage near its mouth (the stage depends primarily on the stream gradient, which varies along the

PRINCIPAL CHARACTERISTICS OF THE STAGES OF STREAM DEVELOPMENT:

Youthful stage

high stream gradient (generally greater than 10 ft/mi)

narrow, V-shaped stream valley

little or no floodplain developed

few, if any, meanders

vertical erosion dominant

Mature stage

moderate gradient (generally less than 10 ft/mi, possibly as low as 1 or 2 ft/mi)

wide, flat-bottomed stream valley with well defined valley walls

floodplain well developed

meanders common; individual meander loops may occupy the full width of the stream valley

transportation and lateral erosion dominant

Old age stage

very low gradient (generally less than 2 ft/mi, often less than 1 ft/mi)

extremely wide valley, perhaps with indistinct valley walls

extensive floodplain, with features such as natural levees

extreme meandering; a distinct meander belt may be developed

deposition dominant

Rejuvenated stage

Tectonic uplift of a region or a lowering of base level may cause the stream gradient to be steepened and an old age or mature stream may be thus rejuvenated. The characteristic feature to look for is the presence of **entrenched meanders** which show that the stream once achieved a low gradient, but that the gradient has since been steepened, reinitiating downcutting. Increased rainfall due to climatic change may also initiate rejuvenation.

PRINCIPAL CHARACTERISTICS OF THE STAGES OF TOPOGRAPHIC EVOLUTION:

Early stage

regional dissection very incomplete, with broad uplands unaffected by erosion

poorly developed drainage system

few streams, mostly in the youthful stage, separated by broad, uneroded interstream divides

local relief **due to erosion** is generally low

Middle stage

regional dissection advanced, few areas unaffected by erosion

well developed drainage system, with maximum number of tributaries

many streams, mostly in the youthful stage, separated by narrow, rounded interstream divides

master streams mature or old age

local relief **due to erosion** is at its maximum development

Late stage

master stream drainage dominates the region, with the master streams in the old age stage

fewer streams than in middle stage due to the merger of stream valleys as interstream divides are completely destroyed by erosion

remaining interstream divides are broad and low

local relief **due to erosion** is once again low, except where monadnocks (erosional remnants) remain

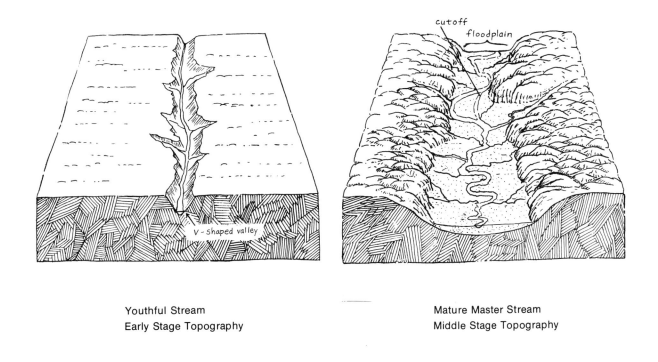

Youthful Stream
Early Stage Topography

Mature Master Stream
Middle Stage Topography

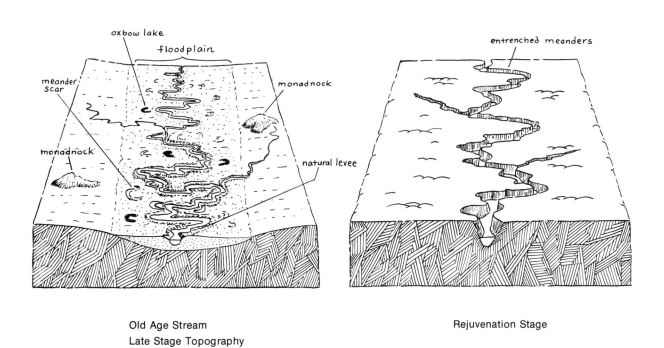

Old Age Stream
Late Stage Topography

Rejuvenation Stage

Fig. 13-1. Stages in the stream cycle and related topographic evolution in humid regions

stream course); and (4) the stages of topographic evolution presented in these diagrams assume a generally humid climate in a region underlain by relatively uniform bedrock. (In later exercises different models of topographic evolution will be presented for more specialized circumstances.).

When studying Figure 13-1, look at each diagram and read the appropriate text on the facing page. After studying each diagram individually this way, try to look at the set as an integrated system.

Looking first at the streams themselves, individual streams or stream segments may be considered to evolve through the stages of youth, maturity, and old age; the primary characteristics of these stages are listed in the left hand column on the page facing Figure 13-1. Note that the main controlling factor in stream development is gradient. Note also that any geologic factor which acts to increase the gradient, such as tectonic uplift, may cause the stream to be **rejuvenated.**

Looking next at the land surface instead of the streams, it is obvious that the development of the stream system will cause the land surface to evolve. In a humid region of relatively uniform bedrock characteristics, the topography will evolve through stages which may be classified as early, middle, and late stage topography, and which have characteristics similar to those outlined in the right hand column on the page facing Figure 13-1.

DELTAS

The sediments carried by a stream system must eventually come to rest somewhere, and the ultimate site of deposition for most stream sediments is the **delta.** A delta will form where a sediment-bearing stream enters a larger body of water such as a lake or an ocean. Although the large mass of water may contain circulating currents, the stream entering this body will nonetheless lose velocity as its waters blend with those of the larger mass. As the stream velocity decreases, the sedimentary load begins to settle out. Deposits accumulate, often building up to the surface of the water to create new land, the visible part of the delta.

Most deltas associated with major rivers are composed mainly of fine-grained sediments — clay, silt, and sand — such as may be carried by low gradient streams in coastal regions. Such a stream, upon entering the delta area, may split up into branches known as **distributaries,** which distribute water and sediments to various portions of the delta.

Several types of deltas are illustrated on page 80.

STREAM SYSTEM PATTERNS

The block diagrams in Figure 13-1 illustrate a semi-randomly branching stream pattern known as dentritic drainage. The type of drainage pattern which develops in a region is easily seen on a map showing the streams, and provides the knowledgeable map reader with considerable insight into the geology of the region. The major drainage patterns are sketched in Figure 13-2 and their geologic significance described.

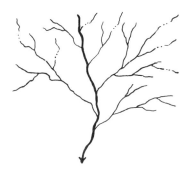

Dendritic drainage: the most commonly developed drainage pattern; develops on rock units which have laterally uniform erosional characteristics, such as flat-lying sedimentary layers or extensive non-structured igneous or metamorphic crystalline rocks.

Trellis drainage: developed in regions underlain by tilted layered rocks such as folded or tilted sedimentary rocks in mountain uplift areas or fold belts; the streams follow valleys eroded in less resistant units such as shale, and are separated by ridges upheld by more resistant units such as sandstone.

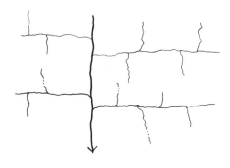

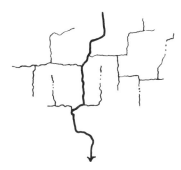

Rectangular drainage: a relatively uncommon drainage pattern which indicates that the bedrock is strongly jointed and/or faulted; the streams tend to follow along the more easily eroded zones, these being the fractures.

Radial drainage: developed around an isolated topographic high, such as a volcanic cone or an area that has been domed up by an intrusion which may reach the surface or may still be buried at depth, depending on how much erosion has taken place.

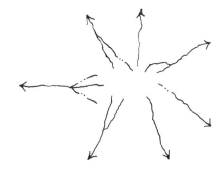

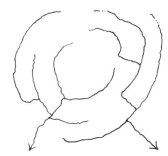

Annular drainage: develops on top of sedimentary strata which have been pushed up into a dome (possibly by an intrusion at depth) and eroded; as in the case of trellis drainage, the streams follow valleys eroded in the less resistant units; as in the case of radial drainage, the streams flow ultimately away from the center of the uplift area.

Centripetal drainage: develops in closed topographic depressions, in arid regions where there is not sufficient rainfall to produce lakes that fill the basin to the point of overflowing, which would produce a through-going drainage; commonly associated with playa lakes in desert basins between mountain ranges.

Fig. 13-2. Stream system patterns.

Development Of Superimposed Drainage

It happens sometimes that a stream course or a regional drainage pattern is developed on a surface underlain by a certain type of rock, but as time passes and erosion takes place, the level of erosion may reach formerly buried rock units that are so different from the rock type on which the streams originally developed that the drainage pattern must change. For example, in an area of complex geologic history, flat-lying sedimentary layers may be unconformably underlain by tilted and folded sedimentary layers. Streams which develop originally on the top of the flat-lying sedimentary layers will have dendritic stream patterns, but when these streams erode down to the level of the folded rocks, the upturned beds will begin to influence the stream courses. The resistant beds will force the smaller, less vigorous streams to turn aside and follow the less resistant beds. Gradually the drainage pattern changes from dendritic to trellis. Only the larger, more powerful streams will be able to erode down through the resistant units and maintain their original stream courses. Such a stream course, developed on one type of bedrock surface and later maintained across a bedrock surface on which this type of stream course would not normally develop, is called a **superimposed** stream course, and the stream itself may be called a superimposed stream.

MAP EXERCISES: In the pages that follow, numerous maps are presented featuring many of the aspects of streams that have been discussed in the preceding pages. The student is asked to study the maps closely and work the problems assigned.

Answer sheets for this exercise begin on Page A-39.

GOVERNMENT SPRINGS, COLO. 1:24000 7½′ C.I. = 20 ft. 1973

1. Calculate the gradient of the stream that flows down Happy Canyon, commencing with the 7800 ft. contour line (just below the primitive road which crosses the canyon near the SW corner of the map) and ending with the 7000 ft. contour line near the last *N* in *CANYON*.

2. Construct a topographic profile from the number *23* in section 23 to the SE corner of section 25, using a vertical exaggeration of 4X.

3. How would you describe the transverse profiles of the stream valleys in this map area? V-shaped? Very wide with indistinct valley walls?

4. Do you see any floodplain features in any of the stream valleys crossing this map area? If so, list them.

5. The streams of this area are intermittent streams, because the climate of the region is semi-arid. Nonetheless, the streams may be classified as youthful, mature, old age, or rejuvenated. How would you classify these streams?

6. Compare the contour map and topographic profile to the diagrams in Figure 13-1 and the accompanying lists of characteristics. What stage of topographic development do you think is represented by this area? (Don't be surprised if the description doesn't exactly fit what you see on the map, for this map area is unusual in that the general land surface has a considerable slope.)

GOVERNMENT SPRINGS,
COLO.
1:24000 7½' C.I.= 20' 1973

LEAVENWORTH, KANS.-MO. 1:62500 15' C.I. = 20 ft. 1948

1. **Sketch** a transverse profile across the Missouri River valley crossing through Sherman Army Airfield.

2. How would you describe this stream valley? V-shaped? Wide, flat-bottomed, with well defined valley walls? Very wide with indistinct valley walls?

3. Sherman Army Airfield is built on a broad flat area which is part of the_____? (Why do you suppose the airfield has a dike or levee built around it?)

4. What is Mud Lake, and how did it form? (Answer in 2 or 3 sentences.)

5. The Missouri River in this area is in which stage of the stream development sequence?

LEAVENWORTH, KANS.-MO.
1:62500 15' C.I.= 20' 1948

CAMPTI, LA. 1:62500 15′ C.I. = 20 ft. 1957

1. How would you describe this stream valley? V-shaped? Flat-bottomed with distinct valley walls? Very wide with indistinct valley walls?

2. Find the lake called "Old River" and the piece of land called "Smith Island"; Old River lake is like Mud Lake on the Leavenworth quadrangle. To prove to yourself that oxbow lakes actually do form in the way you described when answering the question about Mud Lake, **take a dark blue pencil and trace out the old stream course on this map,** starting from Old River lake. (The former stream course is shown by the pair of meandering lines with the dash-dot-dot-dash pattern; these are modern property lines, which were originally based on the stream course, which just goes to show the rapidity with which a stream in this stage of development may wander laterally.)

3. The brown stipple on the inside of many of the meander curves represents deposits of sand called _____ ? (Consult your text or instructor.)

4. Notice that the Red River does not cross a single contour line in this map area; the contour interval is 20 ft.; therefore, it is clear that the river descends less than 20 feet while crossing this map area. Measure the length of the stream course and estimate the maximum gradient in this area.

5. What stage of the stream development sequence is represented by the Red River in Louisiana?

NATCHEZ, MISS.-LA.   1:62500   15′   C.I. = 5 and 20 ft.   1965

This map shows some of the details of the floodplain of a major river in the old age stage of development. Lake St. John is an oxbow lake, cut off from the main channel by a **clay plug.** Several **channel bars** are shown in the river near the NE corner of the map. The Mississippi-Louisiana state line follows an old bend in the river that was abandoned in 1933 when the river eroded a short cut at Giles Cutoff. An intricate network of contour lines near Blue Hole in the south central portion of the map shows a fan-shaped deposit called a **crevasse splay;** this is a deposit of mud formed when floodwaters burst through a **natural levee** (or an artificial one, as appears to be the case here).

Notice the many streamlined and sculptured low relief landforms, all of which reflect deposition and erosion by shifting currents.

NATCHEZ, MISS.-LA.
1:62500 15' C.I.= 5' & 20'
1965

BRETON SOUND, LA.   1:250000   1° X 2°   1957

1. Southwest Pass, South Pass, Southeast Pass, and Pass a Loutre are branches of the Mississippi River system which illustrate the characteristic breaking up of a major river in its delta area. These deltaic branches of the main river are known as _____ ?

2. Notice how each one of the passes is flanked by narrow strips of low land that project out into the Gulf of Mexico. Notice also that the bathymetric contour lines (underwater contour lines) show that the sea floor is higher in the vicinity of the passes. What do you suppose is the cause of these extensions of land and the apparent build up of the sea floor in the vicinity of the passes?

3. Although delta construction by the Mississippi River is the dominant feature in this map area, other processes can be observed here, such as the effects of marine currents. Look carefully at the mouth of South Pass. Here Mississippi waters are deflected by coastal currents. Which way are the coastal currents moving in this area, to the northeast or the southwest? How can you tell?

4. Sketched below are three major types of deltas. Which type is the Mississippi delta?

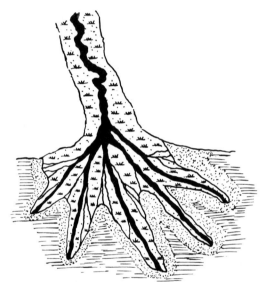

BIRD'S-FOOT DELTA
(Dominated by Stream Deposition)

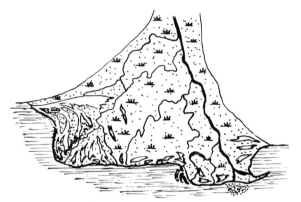

ARCUATE DELTA
(Dominated by Wave Action)

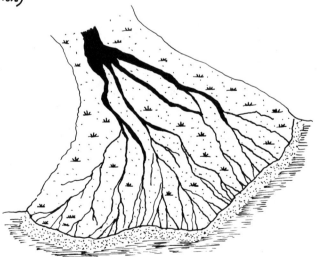

LOBATE DELTA
(Dominated by Stream Deposition)

BRETON SOUND, LA.
1:250000 1° x 2°
Maximum elevation less than
50 ft. above sea level. No
land contour lines.
Bathymetric contours in ft.
below sea level. 1957,
revised 1972

ANDERSON MESA, COLO.   1:24000   7½′   C.I. = 20 ft.   1960

1. The Dolores River (the main river in this map area) reached the late mature or old age stage at one time, but the region has been broadly uplifted and the river gradient was thus steepened, causing the stream to be rejuvenated. How can you tell?

2. Calculate the stream gradient of the Dolores River.

3. Is this gradient typical of a stream with numerous and well developed meanders?

ANDERSON MESA, COLO.
1:24000 7½' C.I. = 20' 1960

HOLLOW SPRINGS, TENN.   1:24000   7½'   C.I. = 10 ft.   1953

1. What drainage pattern is well exhibited by the streams in this map area?

2. Note that the NW half of the map area is highly dissected by an intricately developed network of streams separated by narrow, rounded interstream divides. What stage in the topographic development sequence is represented by this area?

3. In sharp contrast, the SE portion of the map area is an area of much lower local relief, drained by fewer streams, separated by relatively broad, flat interstream divides, some of which show poorly drained swampy areas. What stage in the topographic development sequence is represented by this area?

HOLLOW SPRINGS, TENN.
1:24000 7½' C.I.= 10' 1953

PLUM GROVE, TENN.-VA.   1:24000   7½′   C.I. = 20 ft.   1939

1. What drainage pattern is exhibited by the streams which flow through Stanley Valley? (The name *Stanley* shows in bold type in the valley.)

2. Find Collins Branch and Frozen Branch; these streams and several of the others which drain the hilly area known as the Stanley Knobs exhibit a rather weakly defined **rectangular drainage** pattern. What type of geologic structure is likely responsible for this pattern?

3. What can you say about the rock units in this region as a whole? (Are they likely sedimentary or igneous or metamorphic? Are they flat-lying or have they been tectonically disturbed?)

PLUM GROVE, TENN.-VA.
1:24000 7½' C.I.= 20' 1939

MAVERICK SPRING, WYO.   1:24000   7½'   C.I. = 20 ft.   1951

1. This is an area of semi-arid climate. How can you tell?

2. Trace over all the streams with a dark blue pencil to help bring out the drainage pattern. What type of drainage pattern is this?

3. What kind of rocks do you think crop out[1] in this area? (Massive igneous rocks? Massive metamorphic rocks? Or sedimentary rocks of various lithologies?)

4. What kind of geologic structure is this? (There is an oil field just 1 mile to the NW of this map area; oil accumulations frequently occur in structures like this. See Figure 8-5.)

---

1. "Crop out" is a geologist's way of saying that the rock units are exposed at the surface of the earth.

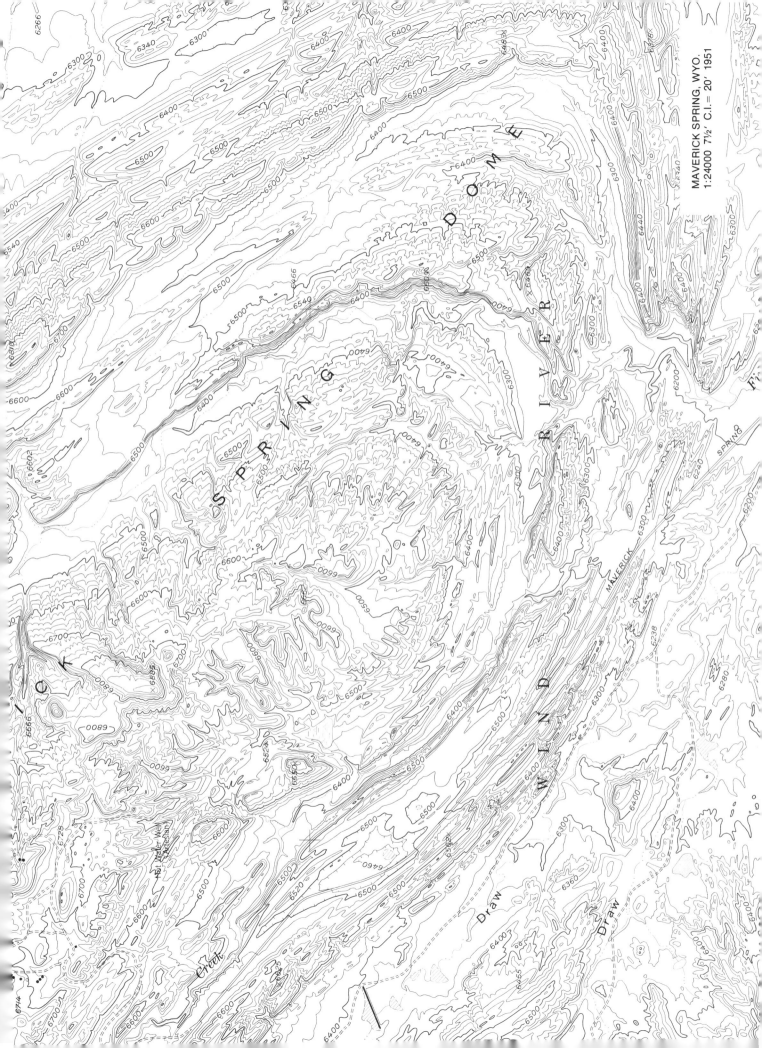

MAVERICK SPRING, WYO.
1:24000 7½' C.I. = 20' 1951

MT. RAINIER, WASH.   1:125000   30′   C.I. = 100 ft.   1924

(Turn to map on page 161.)

   1. The streams descending Mt. Rainier form what sort of drainage pattern?

STRASBURG, VA.   1:62500   15′   C.I. = 40 ft.   1947

(Turn to map on page 58.)

   1. Notice that both the North and South Forks of the Shenandoah River seem to have developed numerous meanders. But are these true meanders? Notice how evenly spaced these "meanders" are, and how straight sided. These curves definitely do not have the normal meander shape. Note also that both streams are high gradient streams (as shown by the numerous rapids); therefore meanders should not develop on these streams under normal circumstances. What geologic explanation can you think of for these peculiar "false meanders"?

HARRISBURG, PA.   1:62500   15′   C.I. = 20 ft.   1956
   (Turn to map on page 66.)

   1) Which is older, the Susquehanna River or the topographic ridge called Blue Mountain? Explain the reasoning behind your answer.

   2) The Susquehanna River is an example of what sort of stream course?

Exercise 14: Karst and Groundwater

Regions underlain by soluble bedrock may be strongly affected by subsurface removal of rock by solutional processes related to ground water movement. Such a region develops underground drainage through cave systems, and a topography characterized by **sinkholes** and other solutional features. If the solutional features become the dominant land form, the topography is called **karst.**

Most of the karst areas of the world occur in regions of humid climate where the bedrock is flat-lying or gently dipping limestone. Solutional features may also be prominent in areas underlain by dolostone, gypsum, or, more rarely, marble or rock salt.

Karst topography is typified by numerous sinkholes, disappearing streams, cave entrances, resurgences (springs), and solution valleys. In a karst region the surface streams will be relatively few and short; through-going streams will be restricted to major streams, or may be non-existent. As the solutional features develop and grow, the topography is modified, and eventually the general land surface is lowered as the soluble bedrock is removed.

Karst features have always been of special interest to man. Caves have served (and still serve) as shelters, places of worship, sources of water, sites of small scale mining of certain minerals such as saltpetre, mushroom farms, natural air conditioning systems for buildings built over them, and as scenic recreational resources. Karst features may also be hazards to human enterprises. Sinkholes sometimes open up abruptly and without warning, engulfing houses and causing loss of property and life. Karst features have also often been abused by man. Sinkholes and caves are frequently used by landowners and municipalities as sites of garbage, trash, and sewage disposal. This is an ill-advised practice which endangers local water supplies and also causes damage to highly specialized and delicate ecological communities which develop in caves.

Figure 14-1 presents block diagrams which represent how karst topography might evolve in a particular area. The stratigraphic sequence shown in these diagrams is (from bottom to top) limestone, shale, limestone, and sandstone. In the **pre-karst stage of development** the area is dominated by normal fluvial erosion as streams flow across the surface of the sandstone caprock. But water passing through joints in the insoluble sandstone has begun to create caves in the upper layer of limestone. The lower layer of limestone is unaffected because the shale acts as an **aquiclude** (a rock layer which will not permit the passage of water).

In the **early stage of karst development,** much of the upper limestone has been exposed by surface erosion of the sandstone unit. Where this limestone is exposed, a **sinkhole plain** has developed. The sinkholes connect to cave systems in the limestone. Surface streams flowing off the remaining sandstone areas become **sinking streams** upon entering the sinkhole plain. Some large caverns have developed in the limestone that is still capped by sandstone, and in places these caverns have collapsed to form **collapse sinks** which break through the sandstone to the surface.

In the **middle stage of karst development,** most of the sandstone unit has been destroyed by a combination of surface erosion and solution in the limestone below it. Remnants of sandstone still cap some of the higher hills. Sinkhole expansion and merger by solution and collapse has produced **solution valleys;** the location and shapes of some of these valleys still reflect the surface drainage pattern that originally developed in the sandstone.

In the **late stage of karst development,** the upper limestone unit has been almost completely removed by solutional processes. A new surface drainage system has begun to form on top of the shale aquiclude. When these streams succeed in eroding through the shale, a new round of karst development will begin in the lower limestone unit.

Answer sheets for this exercise begin on Page A-43

PRE-KARST

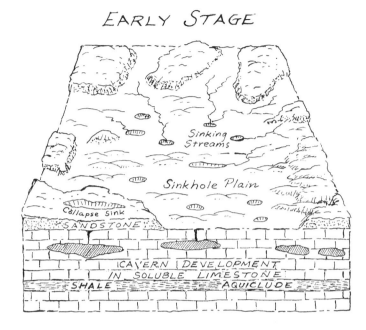

SANDSTONE CAPROCK (JOINTED)
INCIPIENT CAVE DEVELOPMENT
IN SOLUBLE LIMESTONE
SHALE AQUICLUDE

EARLY STAGE

Sinking Streams

Sinkhole Plain

Collapse Sink
SANDSTONE
CAVERN DEVELOPMENT
IN SOLUBLE LIMESTONE
SHALE AQUICLUDE

MIDDLE STAGE

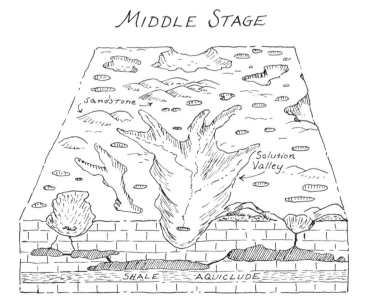

Sandstone

Solution Valley

SHALE AQUICLUDE

LATE STAGE

SHALE AQUICLUDE
L.S.

Fig. 14-1. Karst erosion cycle

MAMMOTH CAVE, KY. 1:62500 15' C.I. = 20 ft. 1922

Three distinct types of topography are shown on this map: the southernmost part of the map is characterized by normal surface streams forming a dendritic drainage pattern; just north of this area is the sinkhole plain, an area of low relief, dotted with hundreds of sinkholes, and lacking normal surface drainage; the northern two-thirds of the map area is characterized by relatively high, flat-topped ridges separated by large solution valleys and dotted with deep sinkholes. **With a red pencil, divide the map area into the three distinct topographic areas.**

The topography in the northern part of the map area is partly controlled by the stratigraphy: a resistant, insoluble sandstone overlies a highly soluble limestone layer. The sandstone layer was partly destroyed by normal surface streams that used to flow across this area. These streams have since disappeared due to the formation of caves and underground drainage in the limestone layer. The sinkholes represent collapse of the sandstone layer into caves in the underlying limestone.

1. What type of bedrock do you think underlies the area of the sinkhole plain south of the high relief area? Is there any sandstone here? Why not?

2. What type of bedrock do you think underlies the area along the southeasternmost part of the map? Why are there no sinkholes here?

3. Notice that the streams flowing northwest across the southernmost part of the map all disappear into sinkholes when they reach the sinkhole plain, e.g., Gardner Creek and Little Sinking Creek. The creek known as Little Sinking Creek disappears into a sinkhole just southeast of the village of Rocky Hill. **Mark this sink on your map with the proper symbol for a cave entrance:** —< The water that goes underground here has been traced with dye to its resurgence point, a group of large springs at Turnhole Bend on the Green River (NW corner of the map area). How many miles (minimum) must the water travel underground before reappearing in these springs?

The underground drainage in this area is quite complex. During times of heavy rain the cave into which Little Sinking Creek normally drains cannot handle all the water, and the creek overflows at the point marked *X* on the map. Here it crosses a field for about a quarter of a mile, and disappears into another sinkhole which is not shown on the map. Water which goes down this sinkhole enters an entirely different cave system, and reappears at a spring in the Barren River about 15 miles to the west!

4. Find Woolsey Hollow, Owens Valley, and Cedar Spring Valley in the area to the SE of Turnhole Bend. These large valleys, called **solution valleys,** are dotted with numerous sinkholes, and the valleys are partly the result of sinkhole growth and merger.

With a blue pencil, draw a line from sinkhole to sinkhole connecting all the sinkholes in the floor of Woolsey Hollow, all the way to Cedar Sink. Next draw a line through all the sinkholes in Owens Valley and connect it to Woolsey Hollow at a sink near the *O* in *Owens*. Finally, draw a line connecting the sinkholes in Cedar Spring Valley to Cedar Sink. What type of normal surface drainage does this branching pattern remind you of? Why do you think the sinkholes line up like this?

MAMMOTH CAVE, KY.
1:62500 15' C.I. = 20' 1922

LOST RIVER AREA, IND.   1:24000   C.I. = 10 ft.   (portions of the French Lick, Georgia, Mitchell, and Paoli quadrangles, Indiana)

Extensive development of karst features can be seen in this area; the map shows numerous sinkholes, disappearing streams, and springs. Especially interesting is the Lost River drainage system.

Lost River is a major stream which disappears underground through a series of sinkholes which have opened up in the stream course. One of these sinks is indicated on the map by the letter *S*. From this point the water travels westward through a cave system, reappearing momentarily in the large "karst window" *(W)* known as Elrod Gulf, and finally reemerging at a large spring or resurgence marked *R* on the map. The minimum underground travel is 3.8 miles.

During times of very high water, the underground drainage system is unable to accommodate the entire volume of Lost River. Instead of going underground, the excess flood waters are forced to follow the old surface channel of Lost River which is normally dry downstream from the sink at *S*. The distance traveled by these surface waters is about 15 miles. **Mark this old stream course with a blue pencil.**

1. During floods, Lost River becomes a two level stream, with an underground branch and a surface branch. Which of these two branches has the greater stream gradient between the points *S* and *R*, Lost River underground, or Lost River surface?

LOST RIVER AREA, IND.
(Portions of the French Lick,
Georgia, Mitchell & Paoli
quadrangles)
1:24000 C.I.= 10'

INTERLACHEN, FLA. 1:62500 15' C.I. = 10 ft. 1949

This map shows a low plain, less than 200 feet above sea level, extensively modified by sinkhole development. The bedrock is limestone of Tertiary age. The water table is relatively near the surface, and is exposed in most of the deeper sinkholes, which are water filled.

1. Note that many of the sinkholes are nearly perfectly round, e.g., Jewel Lake in the east central portion of the map area. Just north of Jewel Lake are two other round lakes, Church and Violet Lakes. Just south of Jewel Lake is another lake, Mirror Lake, which is not round. Why do you think Mirror Lake has this odd shape? (Hint: What will happen to Church and Violet Lakes as solution of the limestone bedrock continues?)

2. The deeper sinkholes form "karst windows" which expose the water table; therefore the depth from the surface to the water table can be estimated by knowing the surface elevation and the elevation of a nearby sinkhole lake. Assume that you have just built a new home near spot elevation 177 in section 33 about 3 miles NNW of the town of Interlachen. Since city water pipes do not reach your house site, you must pay to have a well drilled. The driller claims that he will have to drill at least 160 ft. to hit the water table. Estimate the depth to the water table below your house to see whether you want to let this man drill or find another driller.

3. Estimate the water level in each of the named lakes in the map area. Use these elevations to construct a contour map of the water table underlying the map area. Use a contour interval of 10 ft. (If you do not remember how to contour elevation data, return to Exercise 11. You may also want to consult with your lab assistant before proceeding very far with this project.)

4. Is the water table in this area flat or sloping? If sloping, which way?

INTERLACHEN, FLA.
1:62500   15'   C.I.= 10'   1949

BOTTOMLESS LAKES, N. MEX. 1:24000 7½' C.I. = 10 ft. 1950

The Bottomless Lakes area illustrates a highly unusual type of karst development: gypsum karstification by artesian water.

The Pecos River flows in a valley floored with a thick deposit of alluvium which is underlain by partially eroded strata of the Artesia Formation. This formation consists of gypsiferous beds which form the escarpment and land surface to the east of the Pecos River. Underlying the gypsum-bearing strata is the San Andres Limestone, which crops out to the west of the area shown on this map. Both units dip approximately one degree to the east.

The San Andres Limestone is an **aquifer.** Its recharge area is in the Sacramento Mtns. to the west of the map area. Rainfall soaks into the aquifer in these mountains and travels downdip to the east. Thus the San Andres Limestone to the east of the Sacramento Mtns. is saturated with water; furthermore, this water is under the pressure of a gravity induced "head," due to the weight of the water in the updip portions of the limestone unit.

This "head" or pressure causes the water to move up from the San Andres Limestone into the Artesia Formation. Inasmuch as gypsum is highly soluble, this **upward moving groundwater produces solution features in the Artesia Formation.**

The Pecos River valley is low enough topographically to be below the level of saturation in the San Andres aquifer, hence the upward moving groundwater can reach the surface in this area. The "Bottomless" Lakes are sinkholes 100 to 200 feet deep, formed by solution and collapse of overlying beds into subsurface solution cavities. Many of the sinks are partially filled with water, and several of them used to overflow continuously. Increased demand for water by the inhabitants of New Mexico has resulted in many wells being drilled into the San Andres aquifer. As a result of this increased water usage, the pressure in the aquifer has dropped so much that in 1967 only Lea Lake overflowed, and that water was captured to feed a livestock tank.

Further evidence of upward movement of groundwater in this area can be seen in the form of two springs just south of Lea Lake, and several **spring mounds** marked *M* on the map. These spring mounds are deposits of silt that accumulated around spring orifices.

(The above geologic interpretation adapted from 1967 Christmas card produced by Jas. F. Quinlan.)

1. A geologic **formation** is defined as a rock unit mappable at a scale of 1:25000. Formations are named after some prominent geographic feature in the outcrop area of the rock unit. The Artesia Formation is no exception: it is named after the town of Artesia which lies about 35 miles to the south of the Bottomless Lakes area. Can you think of any geologic reason why the town might be named "Artesia"? Some of the features seen in the Bottomless Lakes area should give you a clue.

2. Find Dimmitt Lake and trace the course of the small intermittent stream which flows into it. Do you think this stream has always flowed into Dimmitt Lake? If not, why not, and where did it flow before it began to flow into Dimmitt Lake? What made it change course? To answer this problem, study the contour lines north of the Fin and Feather Club very carefully, and keep in mind how Dimmitt Lake formed.

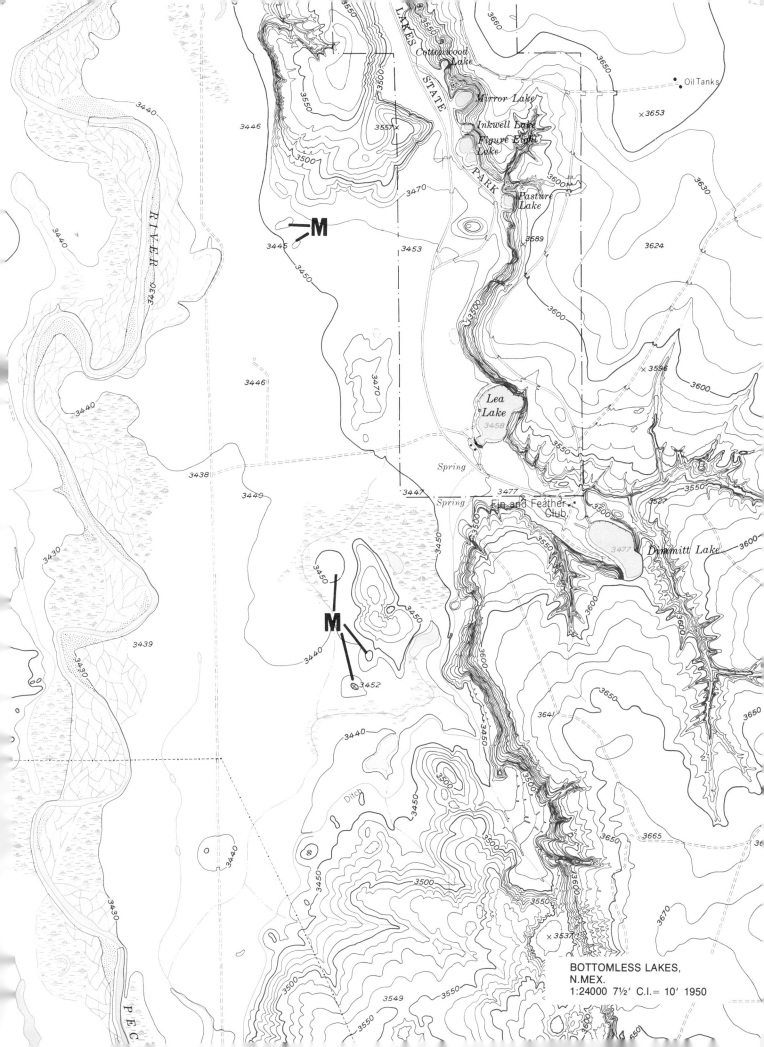

BOTTOMLESS LAKES,
N.MEX.
1:24000 7½' C.I.= 10' 1950

Exercise 15: Topographic Features of Arid Regions

Approximately 30% of the earth's land surface may be classified as desert or semi-desert.[1] Although the climates of these desert areas may be hot, cold, or temperate, they all have one common denominator: annual rainfall is low, water is scarce. As a consequence of this dryness, most deserts are areas of sparse vegetation and thin soils. In such areas, mechanical weathering processes are more obvious than chemical; hence desert topography tends to be sharp and angular. Furthermore, in an area of thin vegetation, the wind may be more effective as an agent of erosion, transportation, and deposition of sediments than it is in humid regions where vegetation tends to prevent wind erosion. Thus it is that many deserts display large areas of bedrock swept barren of topsoil. In other areas, the wind-eroded material may accumulate as sand dunes (Figure 15-2, pg. 118) or loess deposits, but vast areas of sand dunes are not nearly as typical of deserts as most people believe.

Actually, as mentioned in the streams exercise, running water is the most important agent of erosion and deposition, even in desert regions. Fluvial processes progress more slowly in deserts, but they remain the major factor responsible for shaping the land surface.

---

1. This includes the polar deserts, areas which lack moisture due to extreme cold, and which, like other harsh deserts, lack vegetation and soil.

Figure 15-1 shows the effects of fluvial processes on topographic evolution in a special case: fault block mountains in a desert region. The topography illustrated here is typical of the Basin and Range Province of the southwestern United States and northwestern Mexico (the area which includes the Sonoran and Chihuahuan deserts).

In the **early stage** of topographic development, the fault block mountains have been recently uplifted and are not extensively eroded. The mountain scarps are abrupt and relatively straight, reflecting the locations of the faults which form the margins of the uplifted crustal blocks. Streams, which may flow only for short periods after rains, have eroded canyons into the mountain blocks, and the eroded material is deposited in the intermontane basins as **alluvial fans** (the courser sediments) and the **basin fill** (the finer sediments). Because of the lack of large streams, the basins between the uplifted blocks are likely to be **bolsons** with **centripetal drainage** (internal drainage) ending in **playa lakes** which may dry up completely between rains. (Note how geologic terminology from the American southwest is strongly influenced by the Spanish heritage of this region: *bol-SON* is Spanish for *big pocket,* and PLAH-ya means *beach,* which is all that's left when the playa lake dries up.)

In the **middle stage,** the mountains have been significantly dissected by stream erosion. The original sharp fault-line scarps have been destroyed by pedimentation: a **pediment** is a gently sloping erosional surface cut on the bedrock of the mountain block by lateral movements of the eroding streams. (As the streams erode the mountain blocks, they lower their gradients, and lateral erosion becomes progressively more important as the streams approach late youth or early maturity.) In the basin, alluvial fans have grown and coalesced to form **bajadas** (pronounced *bah-HAH-dahs,* Spanish for *descent*), broad alluvial aprons which form gentle slopes descending from the pediment surfaces out into the bolson. The basin fill has continued to accumulate as the mountains are eroded, and the net result is a lower regional relief as topographic development proceeds.

In the **late stage,** the mountain peaks have been reduced to isolated erosional remnants which rise above an extensive alluvium-covered pediment surface. Such peaks are known as **inselbergs.** The regional relief may still be high, but it will be greatly reduced from the original relief present in the early stage.

Answer sheets for this exercise begin on Page A-45.

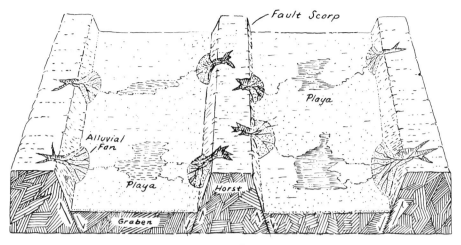

EARLY STAGE

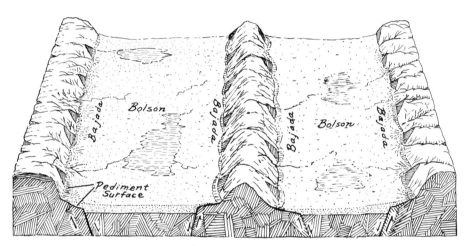

MIDDLE STAGE

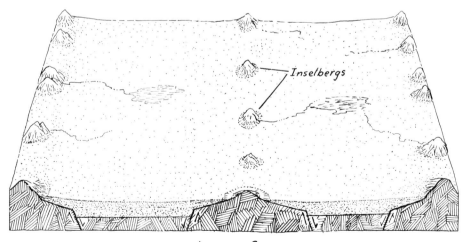

LATE STAGE

Fig. 15-1. Arid region topographic cycle on block-faulted terrain

ENNIS, MONT. 1:62500 15′ C.I. = 40 ft. 1949

This area obviously is not a harsh desert: the mountains are forested, and there is a significant amount of water here. This is a semi-arid region, which shows some features characteristic of arid regions and some features which are also common to humid regions, such as well developed stream terraces.

The Cedar Creek **alluvial fan** is a near perfect example of this particular land form: a large, fan-shaped deposit of sand, gravel, and even boulders, left here as Cedar Creek debouches onto the low gradient valley floor from the high gradient canyon that it has cut into the mountains. The lowering of the stream gradient decreases its competency, forcing the creek to deposit much of the load that it has carried from the mountains.

1. Notice the abrupt topographic break between the valley floor and the mountain front. **Using a red pencil, draw a line on your map, following the sharp break in slope.** This line should show several very straight segments. The mountains have been uplifted along a series of faults, and these straight lines indicate the locations and general trends of the faults. The steep mountain front is known as a **fault line scarp.**

2. What stage of topographic development is this area in, according to the model of topographic development for fault block topography in arid regions? (Early, middle, or late?)

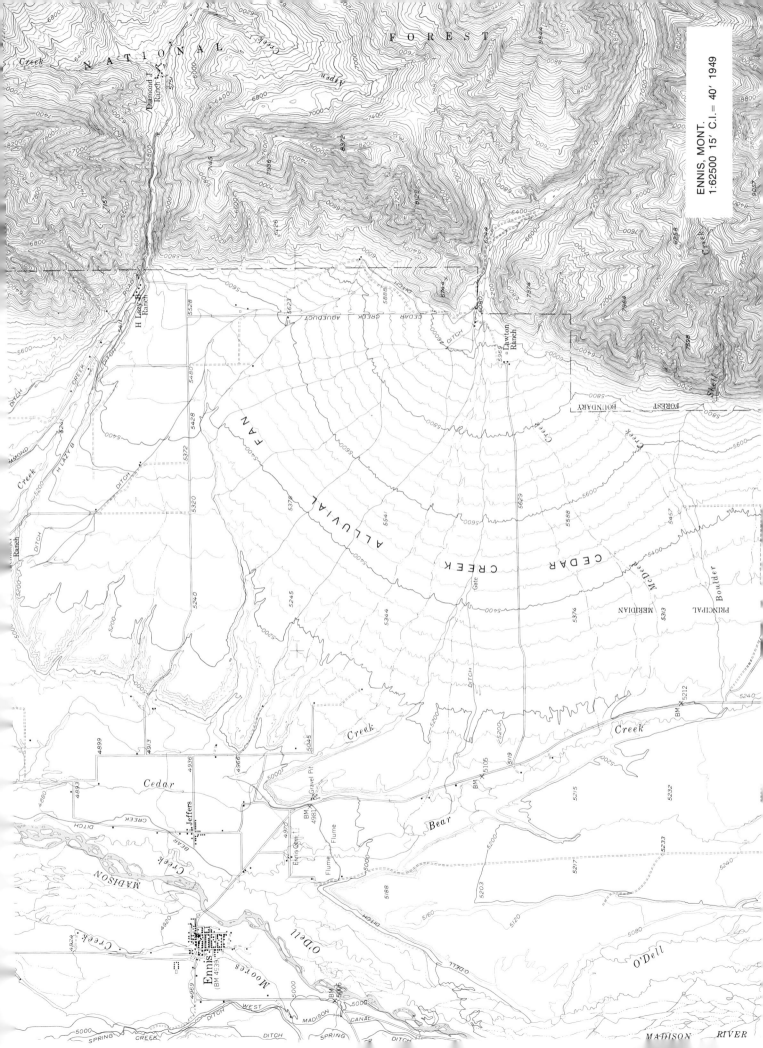

ENNIS, MONT.
1:62500 15' C.I. = 40' 1949

FURNACE CREEK, CALIF.   1:62500   15′   C.I. = 80 ft.   1952

1. Why is the lake shown with ruled lines instead of solid blue?

2. What is the proper geologic name for this type of lake?

3. Near the lake the word *depression* is printed in parentheses, indicating that this is a closed depression, which is not uncommon for desert basins between fault block mountain ranges. But suppose the word *depression* were not printed here. What evidence can you see on the portion of the map that proves this valley is a closed depression? Incidentally, the name of this desert basin is Death Valley.

4. Notice how the alluvial fans in this area have begun to coalesce to form a broad alluvial apron known as a **bajada.** Notice also how the mountain front here is very irregular compared to that shown on the Ennis, Montana, quadrangle. It would be difficult to draw the fault lines here with only the topography to go by; a field examination for other geologic evidence would be necessary. What stage of topographic development is represented by this area?

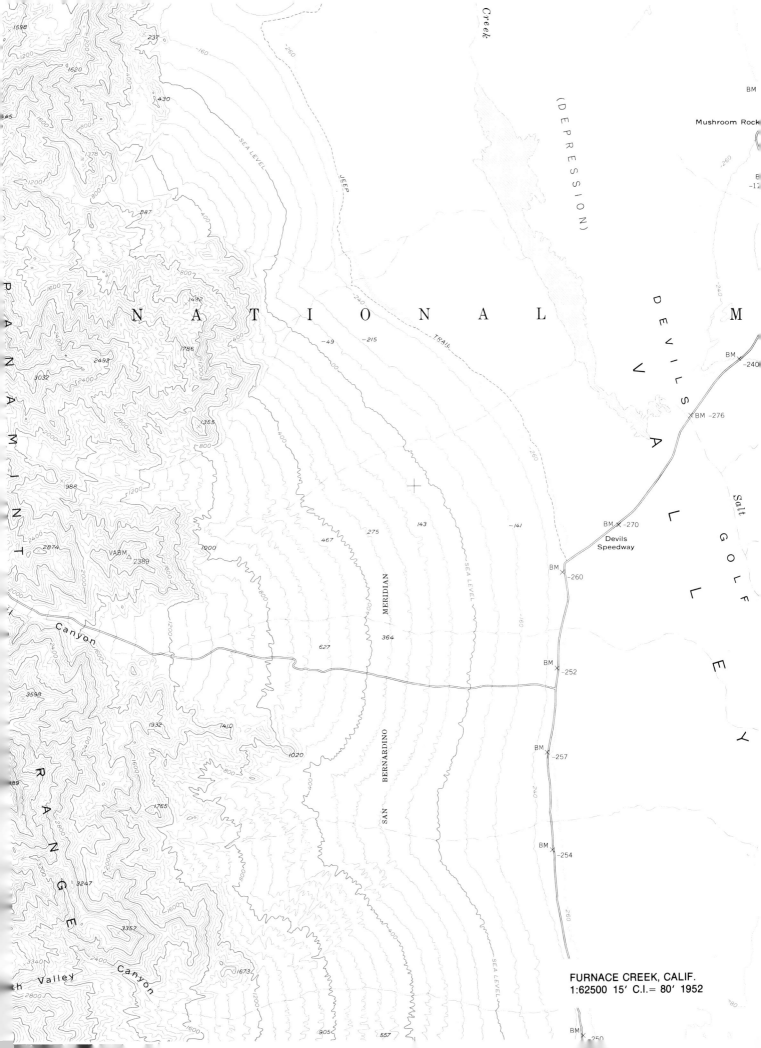

FURNACE CREEK, CALIF.
1:62500 15' C.I.= 80' 1952

ANTELOPE PEAK, ARIZ.   1:62500   15'   C.I. = 25 ft.   1963

In this area it is impossible to tell from the map where the original mountain front may have been. Remnants of peaks, **inselbergs,** rise up from the pediment surface. The extensive pediment is covered with an alluvial sheet and merges at some point, undetectable on the map, with the thicker alluvial deposits of the bajada.

1. What stage of topographic development is illustrated by this map?

2. The pediment/bajada area looks almost flat due to the widely spaced contour lines. Actually the slope is quite gentle, but because it continues for a long distance horizontally, the total vertical relief is considerable. What is the total vertical relief from the SW corner of section 34 in T6S-R2E to the SW corner of section 5 in T6S-R3E?

3. Measure the distance between the two points mentioned above, and with the known vertical relief, calculate the slope between the two points, using the formula given below.

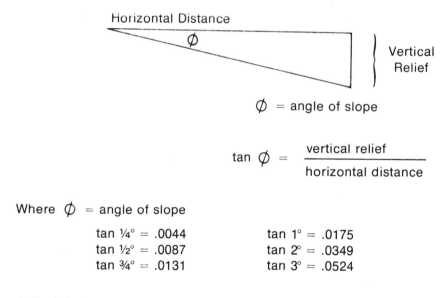

ϕ = angle of slope

$$\tan \phi = \frac{\text{vertical relief}}{\text{horizontal distance}}$$

Where ϕ = angle of slope

tan ¼° = .0044	tan 1° = .0175
tan ½° = .0087	tan 2° = .0349
tan ¾° = .0131	tan 3° = .0524

4. Explain the crinkly nature of the contour lines crossing the alluvial fan just north of Indian Butte. (Note how the crinkles in one contour line are aligned with crinkles in adjacent lines. Look closely at the intermittent stream descending this fan.)

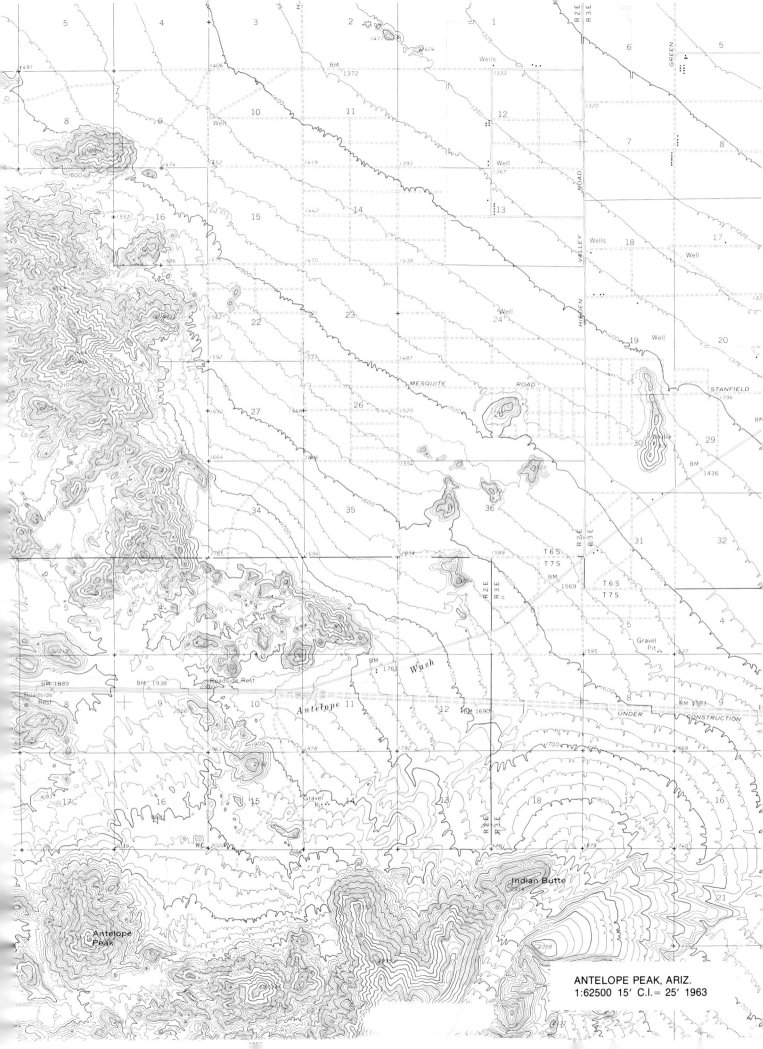

ANTELOPE PEAK, ARIZ.
1:62500 15' C.I.= 25' 1963

NEW HOME, TEX.   1:62500   15′   C.I. = 10 ft.   1957

New Home is a small town on the windswept, semi-arid plains of west Texas known as the Llano Estacado (pronounced "YAH-no Es-tock-AH-do," Spanish for *Staked Plain;* the origin of this name is lost in antiquity).

1. Even if you had no idea what west Texas was like, a few moments study of this map should convince you that it is a dry area. For example, the hundreds of small blue circles are all water tanks, mostly for livestock, showing that water is precious in this region and not abundant in streams. List five other map features which suggest that this is a dry region. (Your list may include features that you see on the map, and also may include features that you don't see that you would see if it weren't a dry region.)

2. This is definitely not a karst area, yet the map shows many dozens of shallow depressions, sometimes called "buffalo wallows" by the local inhabitants. While it is possible that buffalos once came to roll around in these depressions which may hold water or be muddy after a rain, the depressions themselves are definitely not the products of buffalos. What type of erosion do you think might have caused these depressions? (By the way, you should include these depressions in your list of answers for question 1.)

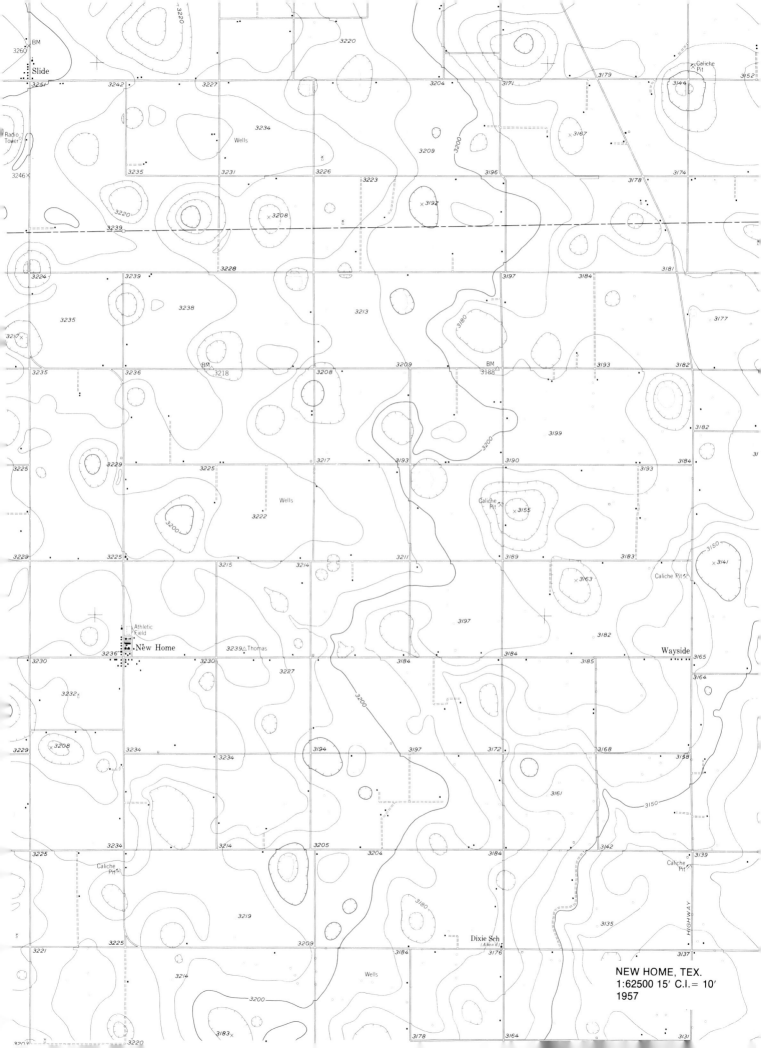

NEW HOME, TEX.
1:62500 15' C.I. = 10'
1957

GLENROCK N.W., WYO. 1:24000 7½′ C.I. = 20 ft. 1949

Like the New Home, Texas, area, this area is strongly affected by wind erosion. Look at the linear features in the topography of this map. This lineation is parallel to the direction of prevailing wind. Now flip back to the New Home map and see if you see a lineation on that map; it's a bit more subtle on the New Home map, but it's there.

1. Although this area is characterized by wind erosion, the wind also deposits. What kind of sand dunes are indicated by the contour lines in the area around the NE corner of section 23? (See Figure 15-2, pg. 118, for dune classification.)

2. Having located and identified these dunes, you should now be able to tell **from** which direction the wind usually blows.

GLEN ROCK N.W., WYO.
1:24000 7½' C.I.= 20' 1949

ASHBY, NEB.   1:62500   15′   C.I. = 20 ft.   1948

Here is an area where the dominant land form is a type of sand dune, yet the region certainly cannot be called a desert, for water is relatively abundant. This should make you wonder if these dunes are relatively old and the climate has become wetter since they were formed.

1. Look very carefully at the minute topography on top of the dunes and decide whether these are **transverse** or **longitudinal dunes.** (Were the dunes deposited crosswise to the wind direction, or parallel to it?) To answer this question, study the dunes in the SW corner of the map area, and also those in the east central part of the map.

2. Now look at the overall shape of the dunes. You should see that most have one gentle slope, the **windward slope,** and one steep slope, the **lee slope.** From which direction did the prevailing wind blow at the time these dunes were deposited?

3. Cite evidence that shows these dunes are stabilized, no longer advancing across this region.

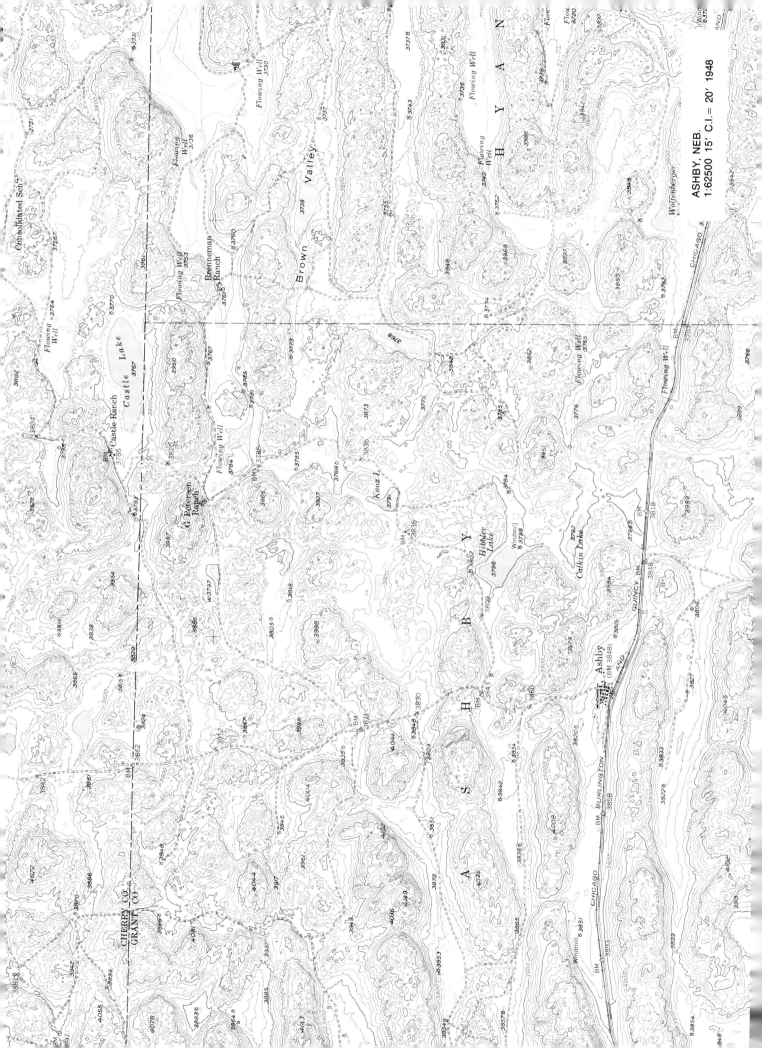

ASHBY, NEB.
1:62500 15' C.I. = 20' 1948

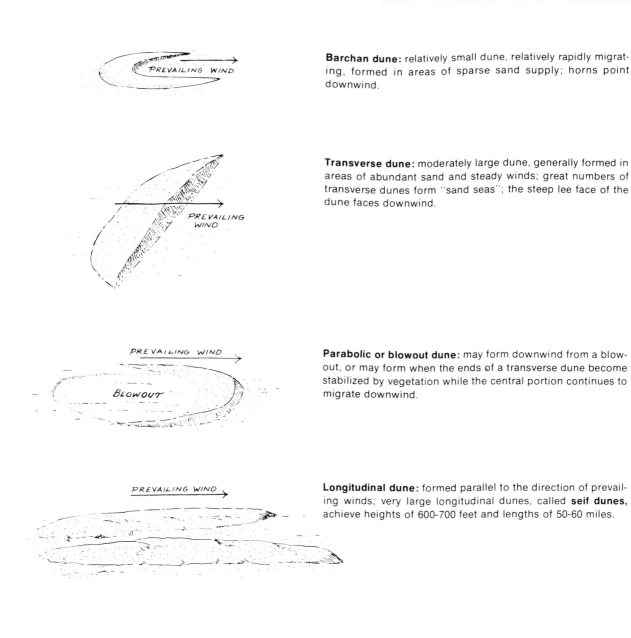

Barchan dune: relatively small dune, relatively rapidly migrating, formed in areas of sparse sand supply; horns point downwind.

Transverse dune: moderately large dune, generally formed in areas of abundant sand and steady winds; great numbers of transverse dunes form "sand seas"; the steep lee face of the dune faces downwind.

Parabolic or blowout dune: may form downwind from a blowout, or may form when the ends of a transverse dune become stabilized by vegetation while the central portion continues to migrate downwind.

Longitudinal dune: formed parallel to the direction of prevailing winds; very large longitudinal dunes, called **seif dunes**, achieve heights of 600-700 feet and lengths of 50-60 miles.

Star dune: irregular shaped dune, generally believed to form in areas of shifting winds, or by vertically moving air columns.

Fig. 15-2. Major sand dune types, classified according to shape

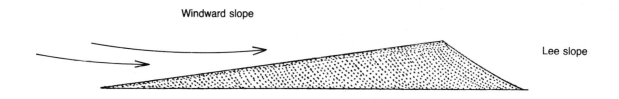

Windward slope

Lee slope

Fig. 15-3. Idealized cross section of barchan, parabolic, and transverse dunes, showing the relationship between dune profile, cross-bedding, and wind direction.

Exercise 16: Alpine Glaciation

Permanent snowfields form in mountainous areas where annual snowfall exceeds annual snowmelt. As snow accumulates, the bottom layers pack to form granular ice known as **firn** or **névé.** This firn consolidates under further compression to form **glacial ice.** When the ice thickness approaches 50m (about 160 ft.) the pressure (weight) of the overlying ice exceeds the bearing strength of the ice on the bottom and this lowermost ice begins to flow plastically out of the **zone of accumulation.** Gradually a "river" of ice forms, flowing slowly down the mountainside following preexisting stream valleys or other natural declivities. An **alpine glacier** is born.

Glaciers are powerful agents of erosion, and the passage of the ice will modify the preexisting stream valleys to produce a distinctive glacial topography that is easily recognized in the field and on topographic maps. Many of these features are represented in the lower diagrams of Figure 16-2.

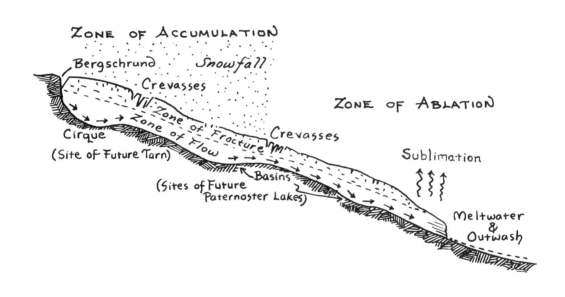

Fig. 16-1. Longitudinal profile of an Alpine glacier

Ice forming below the accumulated snow and firn freezes onto the bedrock surface of the mountainside. As this ice begins to move, it will literally pluck, or pull out, pieces of the bedrock and carry them away. The result of this plucking is the creation of a **cirque,** an amphitheatre-like rock basin partially surrounded by glacially eroded cliffs. After the glacier has melted away the cirque may be partially occupied by a meltwater lake known as a **tarn.**

Where several glaciers flow away from a ridge or mountain peak, several cirques may form. **Headward erosion** by plucking in the cirques will cause the ridge to be narrowed. The peaks may become faceted by cliffs, forming abrupt, jagged **horns** that challenge alpinists. Low spots between horns occur where two cirques begin to intersect due to headward erosion; such saddles are called **cols.** The entire ridge may be transformed into a **comb ridge,** or **arete,** a narrow, jagged ridge of rocky peaks and spires.

The glacier, flowing down the valley, carries large amounts of rock embedded in the ice. These rocks gouge and grind on the bedrock underneath the glacier and in the valley walls. Due to the great weight of the ice mass, these rocks make very effective cutting tools. The result is a widening and deepening of the valley. Characteristically, glacially modified valleys are steep-walled, often rather deep, and flat-bottomed, being generally U-shaped in transverse cross section. At various places along the valley floor, rock basins may be formed where localized plucking and scouring occurs during glaciation. After the glacier has melted, these basins, if separated by relatively abrupt drops in the valley floor, may have the appearance of giant steps and are sometimes called **cyclopean stairs.** The basins will be partially filled with a string of small lakes called **pater noster lakes** due to their resemblance (on a map) to a string of rosary beads.

As the glacier scrapes along the valley walls, rock debris accumulates on the surface of the ice near the margins of the glacier to form **rock trains** known as **lateral moraines.** Where two glaciers flow together the inside lateral moraines merge to form a **medial moraine,** a rock train down the middle of the glacier. These rock train deposits move with the ice and are seldom well preserved as topographic features after the ice has melted. But at the downstream end of the glacier, at the melting snout of the glacier, a larger morainal deposit may accumulate. This **terminal moraine** may make a distinctive topographic feature, but as it is composed of unconsolidated debris these moraines are easily destroyed by later stream erosion.

When melting exceeds the flow rate, the glacial front will retreat, leaving a cover of **ground moraine** on the valley floor. Larger accumulations known as **recessional moraines** occur where the retreating ice front pauses temporarily.

Streams of meltwater issuing from glaciers are often milky colored due to the large amounts of finely ground **rock flour** in the water. These streams redistribute material from the terminal moraine, washing it out into a sloping plain of somewhat stratified and sorted sediments known as the **outwash plain.** Because the melting glacier may release more sediments than the outwash streams are able to move, the stream channels downstream from the outwash plain often become clogged with sediments in the form of numerous channel bars. Thus **braided streams** result from this sedimentary overloading beyond the capacity and competency of the outwash streams.

Answer sheets for this exercise begin on Page A-47.

PREGLACIATION

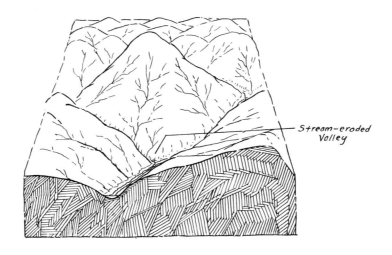

Stream-eroded Valley

GLACIATION

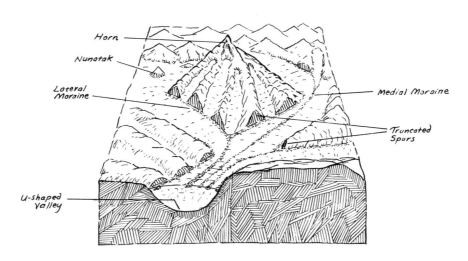

Horn

Nunatak

Lateral Moraine

Medial Moraine

Truncated Spurs

U-shaped Valley

POSTGLACIATION

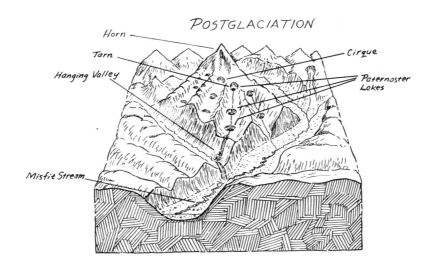

Horn

Tarn

Hanging Valley

Cirque

Paternoster Lakes

Misfit Stream

Fig. 16-2. Topographic modification by Alpine glaciation

McCARTHY, ALASKA   1:250000   1° X 3°   C.I. = 200 ft.   1960

This small-scale map covers a relatively large area and shows a number of alpine glaciers descending from the snow fields of the Wrangell Range. Note the dendritic flow pattern of the glaciers, similar to the stream pattern that would drain this range were the climate warmer.

Study Kennicott Glacier.[1] Notice how small tributary glaciers converge in the mountains above 5000 feet to form the main glacier. Here and there a crag of rock projects up through the ice like Packsaddle Island. The glacier, scraping the valley walls and floor as it descends, incorporates eroded material into the ice and carries it along. Some of this material lies on the surface of the ice, and is shown as brown stipple marks on the map. **Crevasses** form where the ice cracks under strain; these are shown on the map as short blue lines (do not confuse these for the contour lines which are also shown in blue where they cross the ice). Note how the mass of ice acts as a dam across Hidden Creek. Towards the downstream end of Kennicott Glacier the entire surface is marked with stipples, indicating that much glacial debris is accumulating on the surface of the ice. **Wasting** of the ice mass here in the **ablation zone** is accomplished through the processes of melting and sublimation. Downstream from the glacier the stippled pattern extends along the Kenncott River, showing that the river is laden with gravel, silt, and even boulders, nearly all of which have been dumped into the river by the melting snout of the glacier. This sedimentary overloading of the river results in a **braided stream channel:** the volume of water is insufficient to move all the sediments added to the channel, hence the stream breaks up into numerous small channels weaving in and out of sand and gravel bars. This stream pattern is common in areas of active or recently active glaciation.

1. Note that where Regal and Rohn Glaciers join to form Nizina Glacier, a rock train is formed on the surface of Nizina Glacier, downstream from the junction point. Such a rock train out in the middle of a glacier is known as _____ ?

2. Examine the Chitistone River valley. What evidence can you see that suggests that the Chitistone Glacier may have once extended all the way down this valley to its confluence with the Nizina River valley? Cite more than one line of evidence.

3. All factors considered, would you say that the glaciers in this region are advancing or retreating? Cite at least two lines of evidence for your answer.

---

1. The name Kennicott should mean something to you (even if you aren't a geologist at heart). What do you suppose is going on in the area of Bonanza Peak, Porphyry Mountain, and Green Butte? Why is there so much evidence of human activity way out here in the sticks, anyway?

McCARTHY, ALASKA
1:250000 1° x 3° C.I.= 200'
1960

CHIEF MOUNTAIN, MONT. 1:125000 30′ C.I. = 100 ft. 1938

Here is a mountainous area which has been modified by alpine glaciation. Only small remnant glaciers remain, but the topography evinces the former presence of much larger, longer glaciers.

1. Mt. Merritt, Ipasha Peak, Ahern Peak, and Mt. Wilbur are all sharp, high, rugged peaks which have been faceted by glacial erosion, making them roughly triangular in cross section. Such peaks are known as _____ ?

2. Just north of Mt. Wilbur is a very small remnant glacier which occupies a deep rock basin surrounded by semicircular cliffs. Such a basin is a _____ ?

3. Meltwater from the remnant glacier below the peak of Mt. Wilbur forms Iceberg Lake, a type of lake known as a _____ ?

4. Starting at Ipasha Glacier there is a chain of lakes: the lake adjacent to the remnant glacier, Ipasha Lake, Margaret Lake, Mokowanis Lake, Glenns Lake, and Crossley Lake. Notice how the first four lakes in the chain occupy rock basins at distinctly different levels down the glacial valley. Such lakes are known as _____ ?

5. Do you think Chief Mountain itself has been glaciated? Explain your answer.

CHIEF MOUNTAIN, MONT.
1:125000 30' C.I. = 100'
1938

HOLY CROSS, COLO.   1:62500   15′   C.I. = 50 ft.   1949

The Holy Cross region has been modified by alpine glaciation, but not as much as the Chief Mountain area. Notice that while U-shaped glacial valleys are common, there remain fairly extensive areas of mountain slopes which show normal contour spacing. Note that some of the peaks such as Galena Mtn. are not complete horns, being only partially faceted (the SW slope of Galena Mtn. peak is a normal mountain slope, unaffected by glaciation).

1. **Sketch** a topographic profile across the valley of Lake Fork creek. Imagine yourself standing in the valley at the gaging station, looking west, upstream. What would the valley cross section look like? Would it be a V-shaped normal stream valley, or a U-shaped valley?

2. Find Busk Creek, a small tributary to Lake Fork. Note that Busk Creek flows through a small U-shaped valley with a valley bottom elevation between 10500 and 10150 feet above sea level, until it reaches the Lake Fork valley. At the junction of the two valleys, Busk Creek drops abruptly 250 feet to reach the floor level of the larger glacial valley. Busk Creek valley is an example of what type of glacial valley? (See Figure 16-2.)

3. Turquoise Lake is impounded by a natural dam formed by an elongate, semicircular hill that stands about 150 feet higher than the surrounding flat area known as Tennessee Park. This hill effectively blocks the mouth of Lake Fork valley, and dams the creek behind it. What is this hill and why is it here? Is it related to the glacier that once occupied Lake Fork valley?

4. You are correct in supposing that the Arkansas River flows through the state of Arkansas and its waters eventually mingle with those of the Gulf of Mexico. (En route to Arkansas the river also crosses the state of Kansas, and there the people call it the Ar-KAN-sas River!) But what about Homestake Creek in the NW corner of the map? Do the waters of Homestake Creek eventually enter the Gulf of Mexico or the Pacific Ocean? Base your answer on the location of the creek with respect to the **continental divide.**

HOLY CROSS, COLO.
1:62500 15' C.I. = 50' 1949

HAYDEN PEAK, UTAH-WYO. 1:125000   30'   C.I. = 100 ft.   1901

This area has been extensively modified by alpine glaciation; notice how the main ridge of the Uinta Mountains has been reduced to a narrow chain of horns and cols. Two of these prominent horns, Mt. Agassiz and Hayden Peak, were named after prominent nineteenth-century geologists and explorers.

1. The narrow ridge of horns and cols is the result of extensive glacial erosion; such a ridge is known as _____ ?

2. Examine the topography of the major stream valleys. You should see some evidence that quite a long time has passed since the glaciers were active in this area. What do you see that suggests that the glaciation was not very recent?

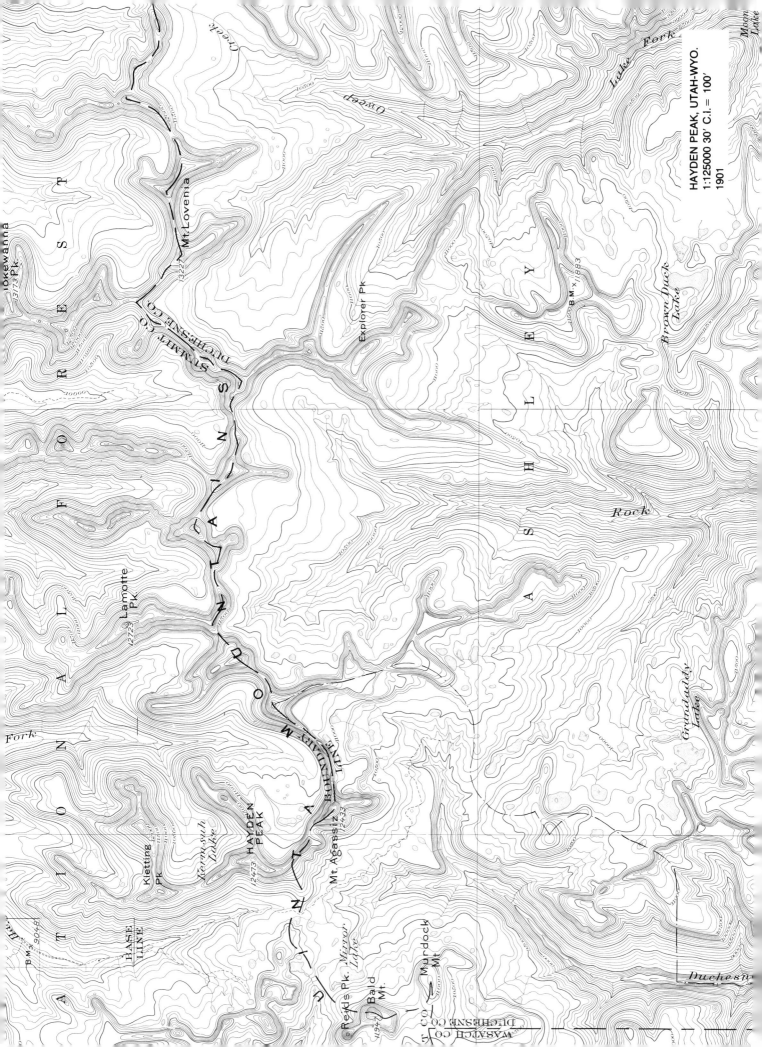

HAYDEN PEAK, UTAH-WYO.
1:125000 30' C.I. = 100'
1901

MT. RAINIER, WASH. 1:125000 30' C.I. = 100 ft. 1924

(Turn to map on page 161.)

Most of Mt. Rainier's slopes are glacier covered in this 1924 map. Do you suppose that this map is still accurate today, more than half a century later? Remember that glaciers are dynamic features. Remember also that we are currently in a period of glacial retreat.

Note the radial pattern of the glaciers which flow in all directions away from the high peak of Mt. Rainier, a prominent volcanic cone.

Note Emmons Glacier, which is named after an outstanding American petrographer.

1. Find "Glacier Island" at about 7 o'clock on the flanks of Mt. Rainier, and "Echo Rock" at about 11 o'clock. Peaks of bedrock which stick up through the flowing glacial ice similar to islands in a river are known as _____ ?

2. Note Ptarmigan Ridge just southwest of Echo Rock. This is a good example of an arête, albeit much smaller than the arête on the Hayden Peak map. Find another example.

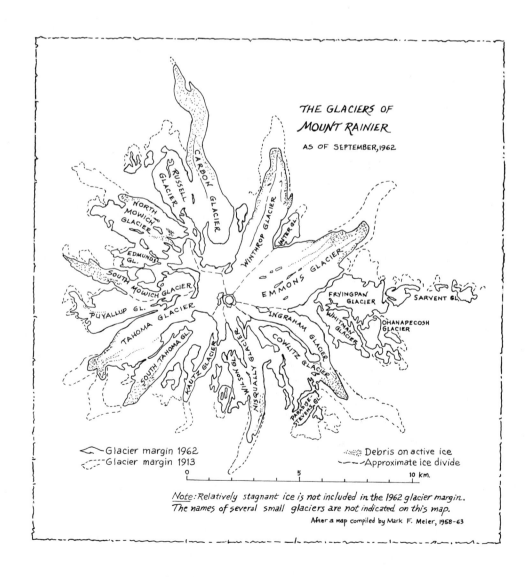

THE GLACIERS OF MOUNT RAINIER
AS OF SEPTEMBER, 1962

Glacier margin 1962
Glacier margin 1913
Debris on active ice
Approximate ice divide

0   5   10 km.

Note: Relatively stagnant ice is not included in the 1962 glacier margin. The names of several small glaciers are not indicated on this map.

After a map compiled by Mark F. Meier, 1958-63

130

Exercise 17: Continental Glaciation

Several times during the history of the earth, climatological and geological factors have combined to produce episodes of widespread **continental glaciation.** These glacial intervals are clearly unusual, and their cause remains a conundrum. But that they have occurred is amply documented by widespread and distinctive glacial deposits and landscapes. The most recent episode of continental glaciation took place in the northern hemisphere during the last million years, and quite likely has not yet ended. During this Pleistocene glaciation the ice sheets grew (advanced) and then melted (retreated) at least four times; stages of glacial advance were separated by interglacial stages. It is quite probable that we are living in an interglacial stage now.

Figure 17-1 illustrates some of the features associated with continental glaciation. In contrast to alpine glaciation (where the paths of the glaciers are controlled by preexisting topographic valleys), in continental glaciation the ice sheet is large enough to override most topographic features. In the region of accumulation the ice sheet may attain thicknesses of more than 3000m (> 10000 ft.). This great thickness of ice forms a topographic high, and the ice flows out in all directions away from this region of accumulation, forming a spreading lensoidal sheet. The weight of this immense sheet may be great enough to actually depress the underlying continental crust; for example, portions of the land in central Greenland are now below sea level, lying under thousands of meters of ice. When the Greenland ice sheet melts away, this land surface will slowly rebound as the land around Hudson Bay is doing now.

As the ice sheet moves across the land, it gouges and scrapes away at the surface, picking up and carrying along vast quantities of debris embedded in the ice. In this way the land surface may be scraped clean of weathered rock and soil; large areas are found in Canada today where bedrock has been stripped clean, striated (scratched and grooved), and polished by Pleistocene ice sheets.

Material eroded and transported by the ice must eventually be deposited. If climatological factors remain constant, the margin of the spreading ice sheet will eventually stabilize in a region where glacial advance is exactly balanced by ablation. A stable front develops, and the ice sheet ceases to enlarge, although the ice flow continues. Debris, deposited by the melting ice, accumulates along the stable front to form the **marginal moraine** (most texts use the term **terminal moraine,** but marginal is used here inasmuch as a continental ice sheet does not have a clearly defined terminus like an alpine glacier). Streams of meltwater issuing from the glacier rework and redeposit some of this unsorted morainal material into a partially sorted, layered **outwash plain.** Occasionally, blocks of ice become detached from the main ice mass and are buried in either the marginal moraine or in the outwash plain. When these blocks eventually melt, the overlying debris caves in and depressions known as **kettle holes** are formed. These depressions often contain small lakes of meltwater. The Minnesota state nickname, ''Land of 10,000 Lakes,'' refers to the abundance of **kettle lakes** in that glaciated region.

Underneath the ice sheet, some very distinctive features may form. **Drumlins** are streamlined hills of **till** (till is a general term for unstratified, unsorted glacial debris). Drumlins are deposits that have been heaped and shaped by the moving ice, and the direction of ice flow may be determined by analyzing the shape and orientation of the drumlins in a drumlin field. A **rôche moutonnée** (sometimes called a **rock drumlin)** is a bedrock knob that has been shaped and streamlined by glacial erosion. Both drumlins and rôches moutonnées are asymmetrical hills, elongate parallel to the direction of ice movement. However, they differ in that drumlins have a steep slope which faces the direction from whence the ice came, whereas rôches moutonnées have a steep slope which faces in the downflow direction. **Eskers** are remarkable, long, sinuous deposits which accumulate on the floors of ice tunnels created by meltwater streams flowing underneath the ice sheet.

During glacial retreat, wholesale melting of the ice sheet may leave vast regions blanketed by **ground moraine.** This irregular deposit, plus changes in the topography due to glacial erosion, may disrupt the normal drainage of a region. Streams in areas covered with ground moraine may have very erratic stream courses; such drainage is called **aimless** or **deranged drainage.**

The term **glacial drift** is applied to all varieties of rock debris deposited in close association with ice sheets.

Answer sheets for this exercise begin on Page A-49.

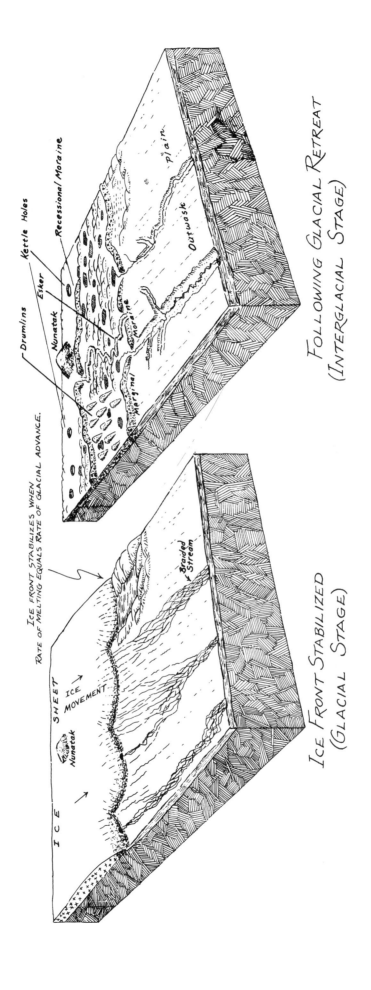

Fig. 17-1. Continental glaciation

Following Glacial Retreat
(Interglacial Stage)

Recessional Moraine

Kettle Holes

Esker

Drumlins

Nunatak

Moraine

Marginal

Plain

Outwash

Ice front stabilizes when
rate of melting equals rate of glacial advance.

Ice Front Stabilized
(Glacial Stage)

ICE SHEET

Nunatak

ICE MOVEMENT

Braided Stream

133

KAATERSKILL, N.Y.   1:62500   15′   C.I. = 20 ft.   1892

This map shows a mountainous area which has been modified by continental glaciation. Note the streamlined topography in the eastern part of the map. Here the smoothed, elongated hills indicate something about the direction of movement of the ice sheet: the long axes of the low hills and ridges are parallel to the direction of ice flow. In the more rugged portion of the map area this "grain" is not apparent, but note that even the high mountains have been affected by the passage of the ice sheet: the mountain topography is smoothed and rounded. This type of mountain physiography contrasts strongly with normal mountain physiography which is more rough and jagged, and mountains which have been modified by alpine glaciation may be even more craggy than normal mountains.

In addition to illustrating mountainous topography modified by continental glaciation, this map shows an interesting phenomenon known as **stream piracy.** Streams which drain the eastern slopes of the mountainous area have gradients which are much higher than those those of the streams which drain generally westward across the mountainous region. For example, note the very high gradients of the small streams descending the east flanks of Overlook and Plattekill Mountains. Then compare these stream valleys to that of the stream labeled "Beaver Kill" in the SW corner of the map. (Budding young conservationists need not to be alarmed; the term *kill* is old-fashioned Dutch New York speech for what folks in other parts of the country might call a "branch" or a "crik"). The streams with the higher gradients will be more competent and more capable of vigorous downcutting than those with lower gradients which expend part of their energy in lateral, sidecutting erosion. Thus it may happen that a high gradient stream, through a combination of downcutting and headward erosion, may cut into the mountain ridge until it intersects one of the lower gradient, westward draining streams. When this happens, the upstream portion of the lower gradient stream will be captured by the higher gradient stream.

Find Echo Lake. Note that water from this lake feeds Saw Kill, which flows more or less straight towards Cooper Lake for about 4 miles before turning abruptly to the south near the community of Shady. Cooper Lake sits at the uphill end of a broad valley that slopes gently away to the west, but the overflow from Cooper Lake drains out to the northeast down a steep gully into Saw Kill. What has happened here is a case of stream piracy. Originally, Echo Lake drained westward into Cooper Lake, which in turn overflowed and drained out the valley to the west. At this time Saw Kill did not extend more than a mile or mile and a half to the north of the area of the present day community of Bearsville. In time, this high gradient stream eroded and extended its valley northward until it intersected the westward flowing lower gradient stream, capturing first the upstream portion of this stream, and later capturing the Cooper Lake drainage.

Historical aside: Note the small town of Woodstock; it was near here that the famous rock festival of that name was held in 1969.

1. What evidence can you see on the map that shows that Cooper Lake used to drain down the valley to the west?

2. What do you think will eventually happen to the headwaters of Beaver Kill? Might another stream piracy take place here? Support your answer with data from the map.

KAATERSKILL, N.Y.
1:62500 15' C.I.= 20' 1892

PALMYRA, N.Y.   1:24000   7½′   C.I. = 10 ft.   1952

The topography of this area is dominated by parallel, elongate, streamlined hills. These hills are composed mainly of unsorted, unstratified glacial debris (**till**) that was pushed, shaped, and overridden by the advancing ice sheet. Clearly the direction of flow was parallel to the long axes of the hills. Furthermore, careful study of these hills will reveal which way the ice sheet moved: the well-formed hills are steeper on the end that faced into the direction of flow, whereas the downstream ends of the hills were dragged out into gently sloping tails.

1. These streamlined hills of till are known as _____ ?

2. Draw a longitudinal profile of the prominent drumlin located between the letters *H* and *E* of *MANCHESTER.* Use a vertical exaggeration of 10X.

3. Which direction did the ice sheet advance across this area, from north to south or from south to north?

4. Which direction did the ice **flow** as the continental ice sheet retreated?

N.B.: Hill Cumorah with its monument to the Angel Moroni is of special interest, for it is here that Joseph Smith reported finding the golden plates that the Angel Moroni had hidden some 1400 years earlier.

PALMYRA, N.Y.
1:24000 7½' C.I.= 10' 1952

JACKSON, MICH.   1:62500   15′   C.I. = 10 ft.   1935

The Jackson, Michigan, area is dotted with hundreds of irregular depressions, many of which are partly water filled. At first glance the topography here looks similar to the karst terrain around Interlachen, Florida (map on page 99). But closer examination of the contour lines on the Jackson map reveals significant differences. The very intricate contour pattern shows that the Jackson area is dotted with many hundreds of very small, very irregular hills and smaller hillocks. Whereas the Interlachen karst area is characterized by depressions in an otherwise fairly regular surface, the Jackson area is knob and kettle topography characterized by odd-shaped hills with depressions left incidentally between hills. The differences between the two areas are fundamental: the Interlachen topography is the result of solutional **erosion,** whereas the Jackson topography is the result of **deposition.** Compare the two map areas until you are certain that you could identify and distinguish between other map areas with the same features.

1. What is the origin of the deposit which covers most of the Jackson map area? What is the geologic name for this deposit?

2. With a dark blue pencil, trace out the stream course of Grand River and all its tributaries. Is this a normal drainage pattern? What sort of drainage is this?

3. Find Blue Ridge on the map. This is a truly extraordinary topographic feature: a discontinuous, sinuous ridge stretching from Skiff Lake near the center of the map area to Schoolhouse Lake near the NE corner of the map area. This peculiar ridge is six miles in length, but less than 1000 feet across and less than 100 feet high. What is the origin of this ridge, and what is the geologic name for such a ridge?

138

JACKSON, MICH.
1:62500 15' C.I.= 10' 1935

WHITEWATER, WIS.   1:62500   15′   C.I. = 20 ft.   1960

Like the Jackson, Michigan, area, the Whitewater area is blanketed with glacial debris left behind after the continental ice sheet retreated from this region. But whereas the Jackson area was mostly covered with ground moraine, the Whitewater area can be subdivided into three areas covered respectively by ground moraine, marginal moraine, and outwash plain.

The area covered by ground moraine has poor natural drainage, but this has been improved by the excavation of small man-made drainage ditches. This area is characterized by small, irregular hills, but in general lacks the depressions seen on the Jackson map. A few of the hills, such as those just east and just north of the town of Whitewater, show some glacial streamlining, but don't have the nice form of true drumlins. Nonetheless, they show that the ice sheet passed over this area, and give some information regarding the direction of movement.

The Kettle Moraine State Forest is a wooded area that has not been converted into farmland, probably because this strip of land is the most irregular topography in this map area. The moraine forms a narrow belt of hills and depressions that crosses the map area from the SW corner to the NE corner. Most of the depressions here are probably kettle holes. In order for this moraine to have accumulated, the ice sheet must have achieved a stable front in this area for some time.

The third distinctive topographic area on this map is the gently sloping surface that extends southeastward from the moraine. This is the outwash plain, composed of stratified, partially sorted glacial debris, deposited by meltwater streams which drained away from the edge of the ice sheet. Some kettle holes occur in the outwash plain.

1. With a red pencil, draw two lines on your map which divide the map area into the three zones discussed above: ground moraine zone, marginal moraine zone, and outwash plain. Label the three zones.

2. Which direction do you think the ice sheet moved as it advanced into this area?

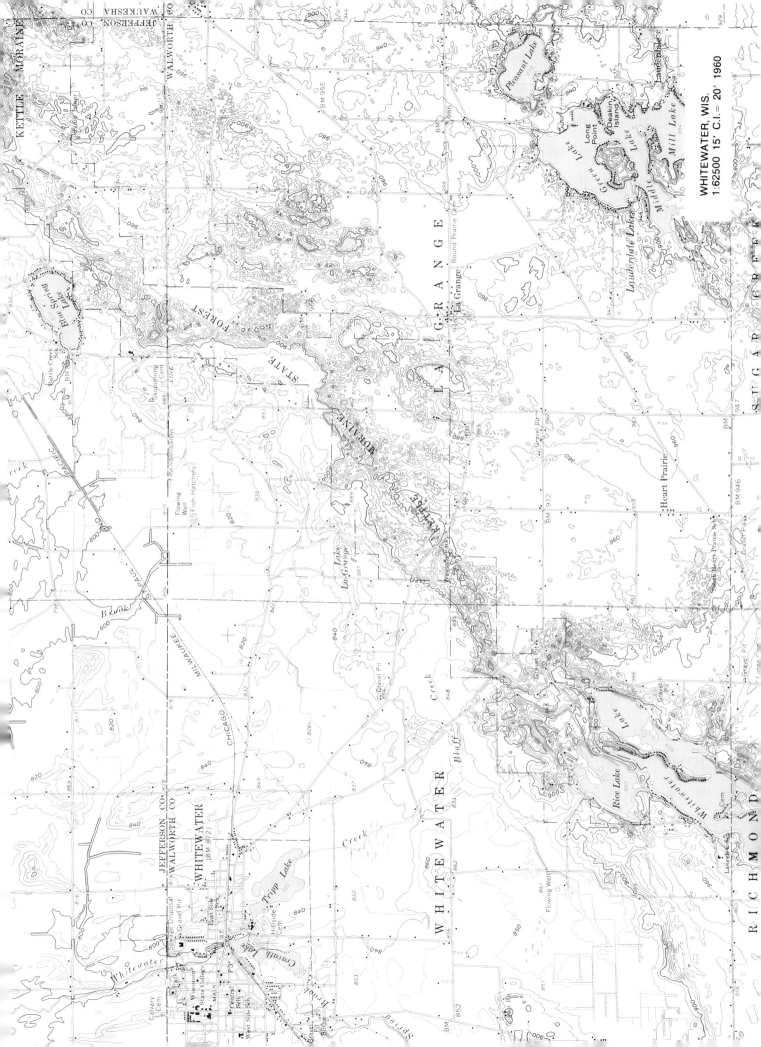

WHITEWATER, WIS.
1:62500 15' C.I.= 20' 1960

BOOTHBAY, MAINE   1:62500   15′   C.I. = 20 ft.   1957

(Turn to map on page 149.)

Much of the coast of Maine has been modified by continental glaciation. Note the streamlining of the ridges in this area. The valleys have been flooded by a post-glacial rise in sea level, due to the melting of vast ice sheets.

1. Do you think the topography here is the result of glacial **erosion** or glacial **deposition?** On what do you base your conclusion?

Exercise 18: Shorelines and Coastal Processes

Shorelines are areas of unusual geologic processes; at the same time, shorelines are zones of increased human endeavor. The unique nature of the shoreline environment responsible for both the unusual geologic and human activity is the fact that the shoreline is a triple point meeting of the hydrosphere (water), the lithosphere (land), and the atmosphere (air). Because this environment is so important to human enterprise, it is especially critical that we understand the geologic processes which shape the shorelines, and control the stability of shoreline features.

The waters of the oceans are always moving, doing work through erosion or deposition. Waves work tirelessly to reshape shorelines. **Longshore currents** work parallel to the shoreline, whereas **density currents** (turbidity, salinity, and cold water currents) move debris perpendicular to the shoreline. Storm waves strip sand from beaches in the winter; the less energetic waves of summer tend to carry sandy debris from the shelf toward the beach. Tides ebb and flow with clocklike regularity, all the while rearranging loose sediment and debris in the nearshore environment.

Most sediment transport in the ocean is directly related to water energies generated by tidal actions, although storms and storm waves can and do generate spectacular changes in shoreline configuration in relatively short periods of time. Storm wave energies can exceed three tons per square foot; storm waves have been known to toss rocks 300 feet into the air, break windows with regularity in lighthouses at heights over 100 feet, and, as at the Tillamook Rock Lighthouse on the Oregon coast, toss a 135-pound boulder through the top of the lightkeeper's house some 100 feet above normal water level.

The more vigorous wave activity tends to occur where the ocean bottom rises rather sharply and wave energy is expended abruptly over a relatively short bottom distance. These conditions accelerate erosion, creating **cliffed headlands, wave-cut benches** or **terraces,** and other features typical of **shorelines of erosion.**

In areas of gently sloping bottom, covered by shallow water, gentle waves carry sand landward to build, with the aid of longshore currents (also wave generated), **bars, spits, tombolos, beaches,** and other features typical of **shorelines of deposition.**

Shoreline configurations are further affected and altered by changes in sea level, both worldwide **(eustatic)** and local **(tectonic).** Interaction of erosion, deposition, and sea level changes creates a variety of topographic forms along the shoreline with simple to complex histories of development. The total effect of waves, currents, storms, deposition, and erosion upon a shoreline depends on various factors, including the geology (rock type, structure, and lithic character); the configuration of the shoreline, the direction, constancy, and maximum velocity of the wind, and the presence or absence of offshore features such as islands, bars, or sea stacks that might deflect waves or currents. This total effect must now also depend significantly upon the activities of mankind in the near-shore environment. Man-made features include sea walls, jetties, breakwaters, and groins (low barriers built perpendicular to the shoreline). Other offshore or shoreline activities such as filling in or farming of tidal flats, laying pipelines and cables, dredging harbors, and industrialization procedures will surely have an increased impact in the future. Population centers near the shore affect geologic processes by increasing runoff (due to the paving of large areas) and concentrating flood waters by channeling. Furthermore, as our population continues to grow, increased recreational use of the shoreline area will inevitably begin to have an effect on natural processes operating there.

Shoreline Classification

Shorelines can generally be classified into four major types: (1) **submergent,** (2) **emergent,** (3) **neutral,** and (4) **compound.** These are distinguishable by shape and depositional or erosional characteristics. The shorelines of submergence and emergence are most common. Shorelines of submergence tend to exhibit many good seaports while shorelines of emergence have few good seaports. Compound shorelines frequently have excellent seaports also.

Shorelines of Submergence

Submerging shorelines occur when a change in sea level results from either a tectonic lowering of the land surface or a world-wide rise in the water level due to melting of large quantities of glacial ice. If the land area is rugged with considerable relief, an irregular coastline with many offshore islands, **estuaries** (drowned river mouths), inlets and bays will result from the submergence. Submerging shorelines of low relief are also irregular and develop estuaries, but in this case **tidal swamps** are also present, the water tends to be shallow, and the shoreline is mostly one of mud and sand.

Wave action on exposed headlands of submerging shorelines will eventually create wave-cut cliffs which recede landward as the rock mass is reduced, forming **cliffed headlands.** Simultaneously, longshore currents and tidal action redistribute the debris at various places along the shore, as shown in Figure 18-1.

Shorelines of Emergence

Emerging shorelines, caused by a relative drop in sea level, will exhibit characteristics of exposed ocean floor. If the original bottom slope was gentle, features such as lagoons, tidal marshes, stranded beaches, and exposed bars will predominate, with the latter two indicating the position of the former shoreline. Emerging shore zones tend to be shallow and generally straight. Incoming waves tend to "break" farther offshore, forming barrier bars that can persist for miles with only an occasional inlet into the extensive lagoon behind. Siltation in these lagoons tends to fill them completely to form new "mainland" after a period of tidal marsh and swampy conditions and the shoreward migration of offshore bars. Much of the Eastern Seaboard of the United States and part of the Gulf Coast falls into this category (see Figure 18-2).

Where emerging submarine topography is steep, with significant relief, deep water conditions will prevent the formation of bars and extensive beaches. Wave-cut terraces, formed between successive periods of emergence, are excellent indicators of this type of shoreline. Examples are found along much of the Pacific Coast of North America.

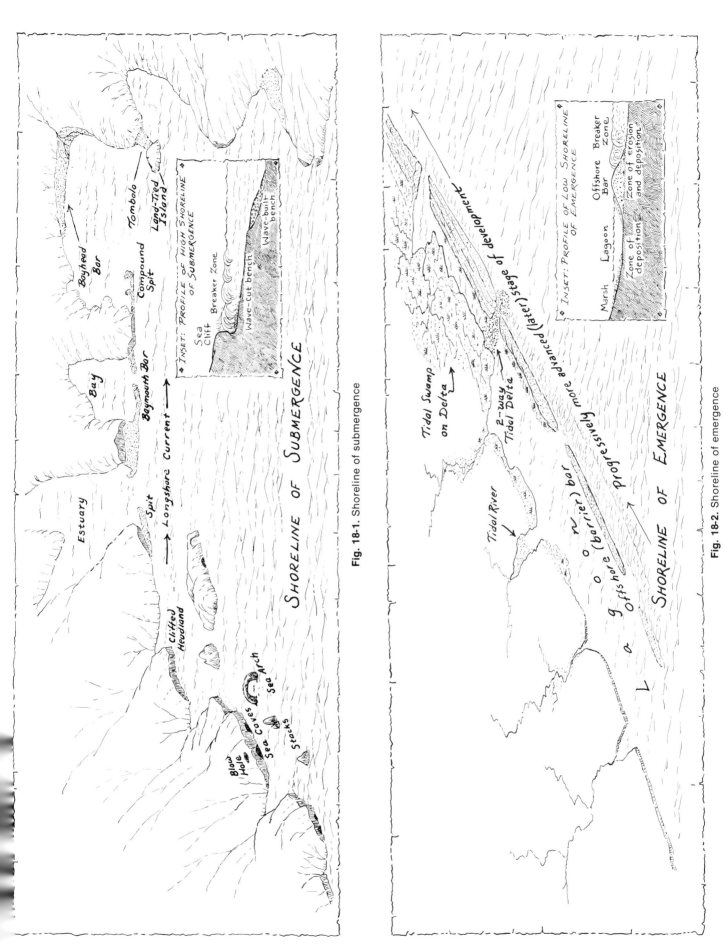

Fig. 18-1. Shoreline of submergence

Shoreline of Submergence

Estuary

Bay

Baymouth Bar

Spit

← Longshore Current

Bayhead Bar

Compound Spit

Tombolo

Land-Tied Island

Cliffed Headland

Sea Arch

Sea Caves

Stacks

Blow Hole

◆ INSET: PROFILE OF HIGH SHORELINE OF SUBMERGENCE

Sea Cliff

Breaker Zone

Wave-cut bench

Wave-built bench

Fig. 18-2. Shoreline of emergence

Shoreline of Emergence

Tidal Swamp on Delta

2-way Tidal Delta

Tidal River

stage of development

progressively more advanced (later)

offshore (barrier) bar

Lagoon

◆ INSET: PROFILE OF LOW SHORELINE OF EMERGENCE

Marsh

Lagoon

Offshore Bar

Breaker Zone

Zone of erosion and deposition

Zone of deposition

Neutral Shorelines

Shorelines which have extended into a body of water by the addition of material from the mainland are termed **neutral.** Materials may be added by volcanoes, rivers building deltas, animal activity, faulting, and development of alluvial fans. The distinction between early and middle stages is not obvious in this type of shoreline, but certainly the early stage is characterized by very rapid outbuilding and the middle stage has begun when waves and currents have smoothed out the irregularities of the material "dumped" during the early stage.

Coral reef shorelines are classed as neutral and are further subdivided into three major types as shown by Figure 18-3. The three types are:

1. Fringing reefs, in which the coral community is attached directly to the shore.

2. Barrier reefs, in which the coral community is separated from the shore by a lagoon.

3. Atolls, reefs which have grown around the perimeter of a subsiding volcanic cone. Continued subsidence and the normal "up growth" of the reef community tend to build a ring of low islands around a central lagoon.

Fringing reefs may evolve into barrier reefs and ultimately into atolls as the volcanic island is eroded or subsides into the sea.

Compound Shorelines

Most shorelines are the result of repeated uplifts and downwarpings of the land surface as well as the outbuilding processes of the neutral shorelines. Combinations of any two of the previously described shoreline types result in a **compound shoreline.**

Answer sheets for this exercise begin on Page A-51.

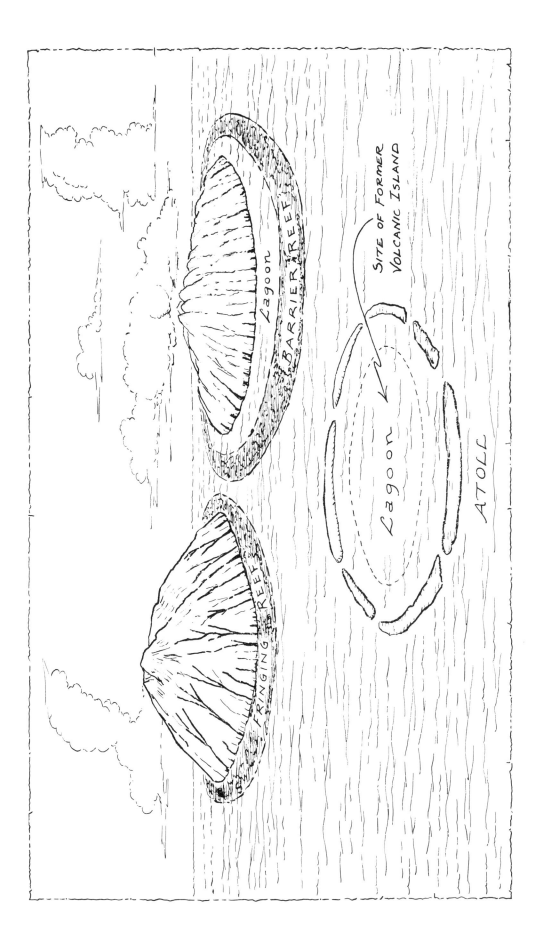

Fig. 18-3. Coral reefs

BOOTHBAY, MAINE   1:62500   15′   C.I. = 20 ft.   1957

1. Is this a shoreline of submergence or emergence?

2. List 3 features that you see on the map that support your answer to question 1. Give the geologic name of the feature and also the map name of the particular example.

3. Give a possible explanation for the steepsideness of channels such as Back River (NC part of map) and Robinhood Cove (WC part of map).

4. Does your answer also account for the parallelism of the ridges and valleys? Explain briefly.

5. Explain the absence of depositional features in this map area. Why don't you find any beaches, bars, spits, or tombolos on the map?

6. Is the water here shallow or relatively deep for near shore water? How can you tell?

7. Find an example of each of the following features, and record the map name of the feature:

   (a) bay   (b) estuary   (c) cove   (d) rocks hazardous to navigation

8. Are good harbors plentiful or scarce in this area?

9. Is the general area safe or dangerous for incoming ships on a very foggy night? Explain briefly. (Hint: Note the unusual map symbol in Mill Cove by the town of Boothbay Harbor.)

BOOTHBAY, MAINE
1:62500 15' C.I.= 20' 1957

LYNN, MASS.   1:24000   7½′   C.I. = 10 ft.   1956

1. Is this a shoreline of submergence or emergence? List three map features that support your answer.

2. Locate a tombolo on the map. Is it a complex or simple tombolo?

3. Identify a bayhead beach.

4. On the north side of Nahant Island is a place marked *Johns Peril.* What do you suppose is the nature of the peril here?

5. From which direction does the strongest wave action come? List evidence to support your answer.

6. Compare the relative depth of water east and west of Little Nahant Island. Explain why the water is distinctly shallower on one side than on the other.

7. What do the dashed blue lines indicate?

Outer Gu

Dread Ledge

Egg Rock

N A H A N T   B A Y

Little Nahant

Little Nahant
Beach

U. S. Coast Guard
Station

Sand

Nahant Beach

Parking Area

BEACH   PARKWAY

NAHANT

Sand Pt

Sand

Sand

Lynn

Light

Light

Light

Light

Light

Light

Light

LYNN   HARBOR

Sand

Sand

Sand

Powerplant

Drive-in
Theater

General Edwards
Bridge

Roosevelt
Sch

Point of Pines

BM

Sand

Sand

WESTERN   CHANNEL

SUFFOLK CO
ESSEX CO

LYNN, MASS.
1:24000 7½' C.I. = 10' 1956

BLACK ROCK CHANNEL

Black Rock
Beach

Black Rock Pt

Lobster
Rocks

West Cliff

Bass Pt

Lewis
Cove

Baileys
Hill

Johnson
Sch

Jackson
Park

Dorothy
Cove

N A H A N T

Nahant

Nahant Harbor

Willow Rd

POND

NAHANT

Valley Road
Sch

Greenlawn
Cem

Wilson
Sch

Johns Peril

Stony Beach

Black Mine

Spouting Horn

Saunders
Ledge

Forty Steps

Cedar Pt

Library

Town Hall

CHURCH ST

VERNON ST

ROAD

ROAD

Josephs Beach

Bass
Rock

Swallow Cave

Pea Island

Northeastern Univ
Edwards Laboratory

East Pt

Great Ledge

Shag Rocks

POINT REYES, CALIF.   1:62500   15′   C.I. = 80 ft.   1954

1. What is the scale of this map? The contour interval?

2. Is the relief greatest in the eastern or western half of the map?

3. What is likely to happen in the future to Drakes Estero and Estero de Limantour? (*Estero* is Spanish for *estuary*.)

4. Give the map name of an example of each of the following features:

   (a) spit   (b) cliffed headland   (c) stack
   (d) cove   (e) estuary   (f) beach

5. Note the bathymetric contour lines. Would you say that the ocean floor sloped off more steeply in Drakes Bay than it does immediately south of Point Reyes, or vice versa?

6. Pick the best harbor site in this map area, and explain why you picked it.

7. Is there a longshore current operating in Drakes Bay? If your answer is yes, which direction does this current flow?

POINT REYES, CALIF.
1:62500 15' C.I. = 80' 1954

BEAUFORT, N.C.   1:250000   1° X 2°   C.I. = 50 ft.   1972

1. What is the scale of this map? The contour interval?

2. Is this a shoreline of submergence or emergence? Explain your reasoning.

3. Give the map name of an example of each of the following features:

    (a) lagoon   (b) barrier bar   (c) estuary   (d) inlet

4. What does the future hold for Core Sound? Will there always be water here? Explain your answer.

5. Is this an area of high relief or low relief? Cite reasons for your answer.

6. Will the gradients of the streams in this area be high or low?

7. Does the water in Adams Creek Canal flow the same direction all day long? Why not?

BEAUFORT, N.C.
1:250000 1° x 2° C.I.= 50'
1972

SAN CLEMENTE ISLAND CENTRAL, CALIF.   1:24000   7½′   C.I. = 25 ft.   1943

1. Using the graph paper provided in the answer sheets, construct a topographic profile along line A-A′. Use a vertical exaggeration of 2X. At the SW end of the profile line, you will have to sketch in the underwater topography in what seems to you to be geologically reasonable when the rest of the profile is examined.

2. What is the origin of the alternating steep, cliff-like slopes and relatively flat terraces which show on both the map and topographic profile?

3. Do these map features indicate that a relative change in sea level has occurred in this area? If so, did sea level rise or fall? How much?

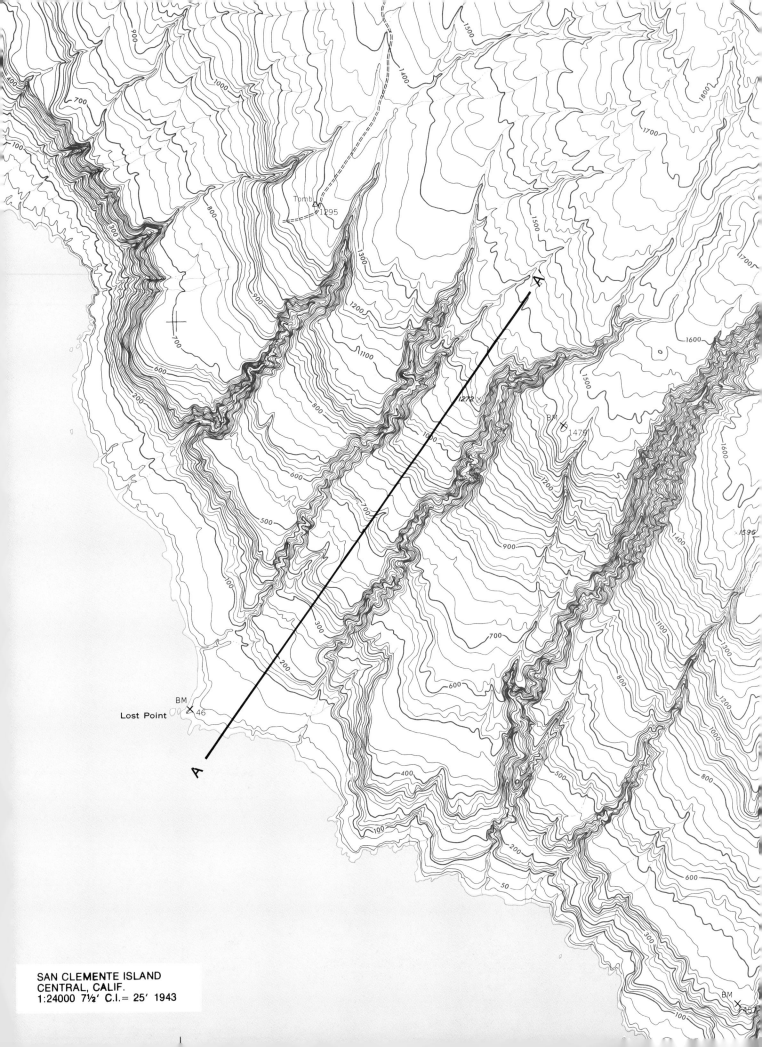

Lost Point

A

A'

SAN CLEMENTE ISLAND
CENTRAL, CALIF.
1:24000 7½′ C.I.= 25′ 1943

Exercise 19: Volcanic Landforms

Volcanic eruptions are responsible for some of the most interesting and spectacular topographic features on earth. Since most volcanic landforms are caused by a buildup of volcanic rocks on the surface, these processes and landforms may be considered **constructional** (in contrast to most landforms caused by erosion, which are **destructional).**

Three basic types of volcanoes are recognized: cinder cones, shield volcanoes, and composite cones, as sketched in Fig. 19-1.

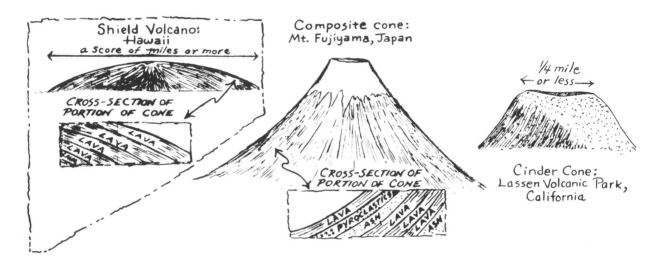

Fig. 19-1. Shapes of volcanic cones

Cinder cones are relatively small, generally not more than a few tens to a few hundred meters high. They are the result of explosive ejection of cinders, ash, and other pyroclastic material. This material is blown into the atmosphere where it cools and congeals before falling back to earth. The ejecta accumulates around the vent, gradually building up a steep-sided cone of relatively loose pyroclastic material.

Shield volcanoes are the result of repeated eruptions of lava flows which spread out from the vent area. The cumulative result of numerous eruptions is a broad-based mountain much wider than tall, with a convex upward profile (shield shaped). Some of the largest mountains on the earth are shield volcanoes, such as the volcanoes which rise over 8500m (28,000 ft.) from the sea floor to form the Hawaiian Islands. Most shield volcanoes are composed primarily of basaltic flows.

The well known scenic volcanoes of the world, such as Mt. Fuji, Mt. Mayón, Volcán Agua, and Popocatepetl, are **composite cones.** These beautiful, soaring cones are composed of flows interlayered with pyroclastic ejecta. The flows spread out from the vent, whereas the pyroclastic ejecta accumulates near the vent; the resulting cone has slopes which steepen towards the vent. In size, these cones fall between the other two cone types; composite cones may form mountains several thousand meters high, but they are neither as broad nor as high as the really big shield volcanoes.

Other volcanic features will be discussed in conjunction with the maps.

Answer sheets for this exercise begin on Page A-55.

MENAN BUTTES, IDA. 1:24000 7½' C.I. = 10 ft. 1951

(Turn to map on page 41.)

1. Construct an east-west topographic profile across North Menan Butte, using a vertical exaggeration of 4X.

2. How tall is North Menan Butte? Calculate the total vertical relief from the highest marked peak down to the Henry's Fork of the Snake River.

3. What type of volcanic cones are the Menan Buttes: cinder, composite, or shield? Compare your profile to the sketches in Figure 19-1 and see which it matches best.

Note the peculiar topography to the north of the Menan Buttes. This area is covered with basaltic flow rocks which probably issued from **fissure vents,** possibly connected to the same magma chamber which fed the Menan Butte cones. What appears on the map to be "sinkholes" are depressions in the lava flow surface, some caused by collapse of the lava crust into voids left behind by migrating molten lava.

MT. RAINIER, WASH. 1:125000 30' C.I. = 100 ft. 1924

1. What evidence can you see on the map that Mt. Rainier is a volcano? (Hints: Look for the crater. What about the drainage pattern?)

2. From the map, does it appear to be an active or inactive volcano? Explain your answer.

3. How tall is Mt. Rainier? Calculate the total vertical relief from the highest marked point on the peak down to the Superintendent's Headquarters on the road that enters the National Park from the southwest.

4. What type of volcanic cone is Mt. Rainier: cinder, composite, or shield? (Hint: Do you think cinder cones get this big? Are shield volcanoes this steep?)

MT. RAINIER, WASH.
1:125000 30' C.I.= 100'
1924

MAUNA LOA, HAWAII   1:62500   15′   C.I. = 50 ft.   1928

This map shows a very tiny portion of the entire volcanic mountain. Note the even spacing of the contour lines, which shows that the slopes down from the main craters are very even. Note how the various craters and vents form a NE trending line; the dashed brown lines representing fractures also follow this trend. Evidently there is a major fracture system through the mountain which controls the location of the vents.

1. What is the most recent volcanic activity indicated on this map?

2. Calculate the total vertical relief from the highest point shown on the map to the lowest index (dark) contour line in the SE corner of the map area. (Remember, only a small portion of the overall volcano is shown on this map.)

3. What type of volcanic cone is Mauna Loa: cinder, composite, or shield?

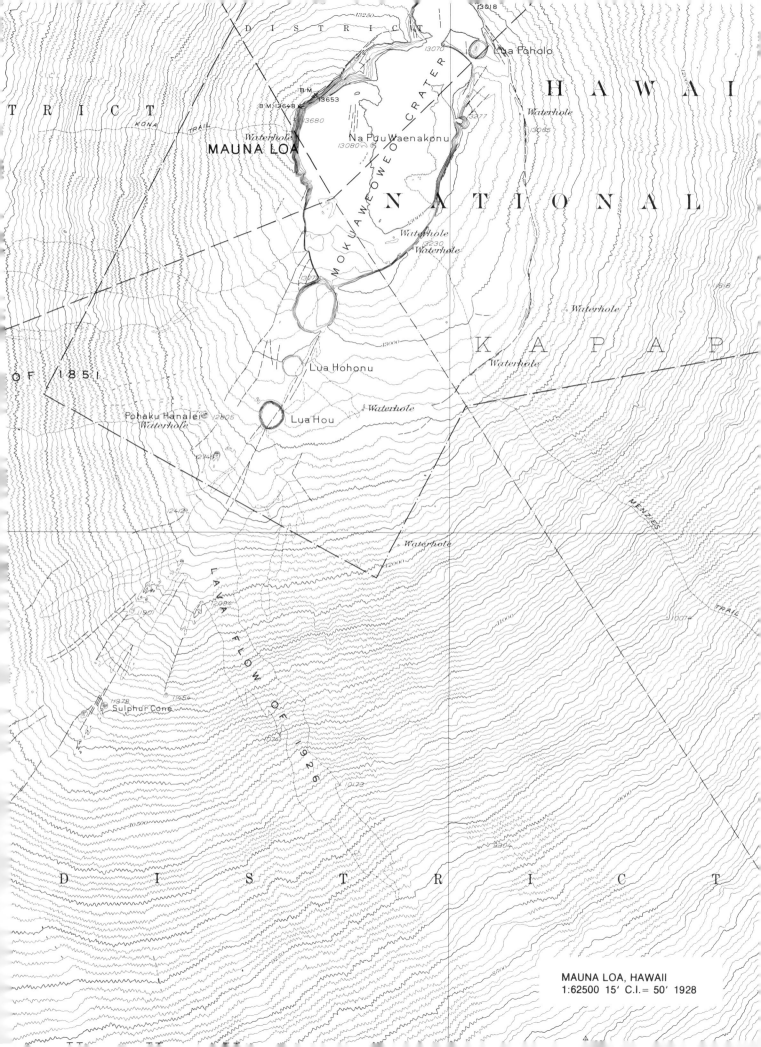

D I S T R I C T

T R I C T

KONA   TRAIL

BM
BM 13648   13653
13680

MAUNA LOA   Na Puu Waenakonu
13080

Waterhole

OF   1851

Pohaku Hanalei   2805
Waterhole

2748

2473

1901   2097

1973   1554
Sulphur Cone

10500

Lua Poholo

HAWAI

Waterhole
13277
13085

MOKUAWEOWEO CRATER

NATIONAL

13000

Waterhole
13230
Waterhole

13277

Lua Hohonu

Lua Hou   Waterhole

Waterhole

Waterhole

KAPAP

1618

Waterhole

Waterhole

12000

Waterhole
12000

MENZIES

11000
1007

1024

1173

TRAIL

9000

9304

850

D   I   S   T   R   I   C   T

13018

13070

12500

LAVA FLOW OF 1926

MAUNA LOA, HAWAII
1:62500 15' C.I.= 50' 1928

MT. DOME, CALIF.-ORE. 1:62500 15' C.I. = 40 ft. 1950

1. Three prominent scarps, ranging in height from a few tens of feet to over 400 feet, cross this map area trending almost due north. Construct an east-west topographic profile across these three scarps, passing through Howitzer Point. Use a vertical scale of 1" = 800 ft.

2. Assume that there is only one geologic rock unit exposed along the line of your profile and that the topography is the direct result of faulting. What type of faults do these scarps most likely represent: normal, reverse, thrust, or strike-slip?

Note the small ponds formed in depressions at the base of some of the fault scarps. These depressions are on the down-thrown fault block. Small ponds found in depressions along fault lines are known as **sag ponds** and are sometimes very helpful to the field geologist as a clue to the location of a fault zone that may not have any obvious scarp or other topographic features.

3. The southern portion of the map area is covered by relatively recent basalt flow rock. It is not clear from the map whether all the recent lava is due to a single flow or several; for simplicity's sake, assume that all the recent basalt indicated by the irregularly patterned surface is due to a single flow. **Outline this flow on your map, using a red pencil.**

4. Study the relationship between the fault scarps and the distribution of the recent flow. Do you think the faulting has occurred since the eruption of the basalt flow? Or were the fault scarps there at the time of the basalt eruption?

Historical note: Captain Jacks Stronghold (east central border of map area) was the Modoc Indian stronghold during the "Modoc War" of 1872-3. Here a band of about 50 Modoc warriors, led by "Captain Jack," held at bay for fully three months U.S. Army troops which eventually numbered 800. The Indians were aided in their tenacious fight to avoid capture and removal to a reservation by the very rugged nature of the lava fields in which they made their stand. The Modoc War cost the U.S. Government 160 dead white citizens, 16 dead Indian citizens, and an estimated $1,000,000. The government's goal of moving the Indians onto a reservation was achieved. Ironically, the land that the Indians wanted to remain on could have been purchased for about $20,000. (Data from "Modoc Indian War" by H.V. Sproull, 1975, Lava Beds Natural History Association.)

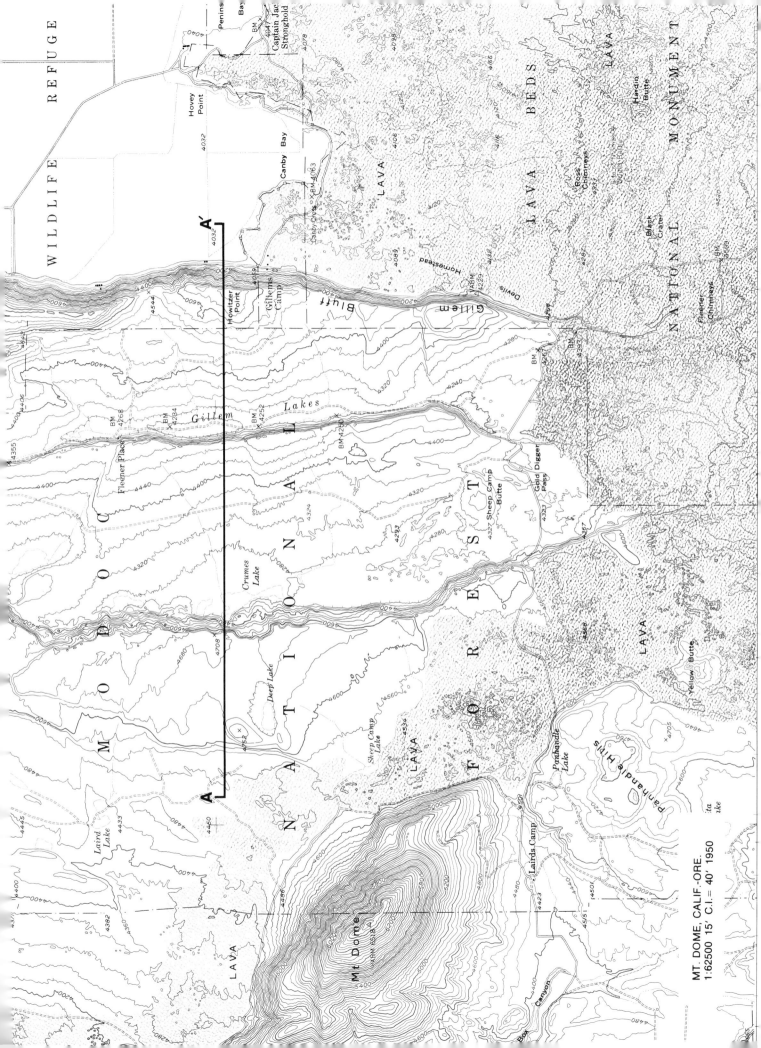

MT. DOME, CALIF.-ORE.
1:62500 15' C.I.= 40' 1950

CRATER LAKE NATIONAL PARK AND VICINITY, ORE.   1:62500   C.I. = 50 ft.   1956

Crater Lake has the distinction of being the sixth deepest lake in the world, which is rather amazing when you consider that it is situated at the top of a mountain. No streams flow into the lake; the waters are entirely derived from rain and snow, and as a result, the water is very pure and clear. The lake is a dazzling cobalt blue, and the park is without doubt one of the most beautiful in the world.

It was not always thus. Prior to the formation of the **caldera** now occupied by Crater Lake, a lofty volcanic peak, which geologists now refer to as Mt. Mazama, stood on this site. By studying the profile of the present-day mountain it has been estimated that Mt. Mazama was about 4000 feet higher than the present caldera rim. About 6600 years ago (based on radiocarbon analyses) a series of violent eruptions took place and Mt. Mazama discharged an estimated 12 cubic miles of magma in the form of hot ash and pumice, which rained down over an area of more than 350,000 square miles. Following these mighty eruptions, which were undoubtedly awe-inspiring to the Indian inhabitants of the region, the evacuated magma chamber collapsed and the top of the mountain caved in to form the caldera which is now occupied by the lake. A **caldera** is a crater-like volcanic structure which is larger than the original, true eruptive crater, and which results from violent explosion and collapse of the original volcanic cone. Thus the name *Crater Lake* is something of a misnomer, but it does have a better ring than *Caldera Lake* so the geologic community has agreed not to lobby for a rechristening!

Sometime after the main eruptions, smaller, less violent eruptions occurred within the caldera, and several small parasitic cinder cones were built up on the caldera floor. One of these cinder cones stands above the lake level today, forming Wizard Island. Another is indicated by the bathymetric contours southeast of Pumice Point on the north side of the caldera; this submerged cone is known as Merriam Cone.

1. What is the greatest water depth of Crater Lake as actually shown on the map?

2. How deep is the water over the top of Merriam Cone? Note that the floor of the lake must rise very abruptly from the 1800 foot depth contour to the top of Merriam Cone, thus the flanks of the cone must be quite steep. This suggests that the cone formed before the lake, inasmuch as sub-aqueous eruption would build a more gently sloping cone.

Note: It is recommended that the lab instructor provide the student with the complete Crater Lake National Park and Vicinity, Oregon, quadrangle. The geologic history of Mt. Mazama and Crater Lake, as interpreted by Dr. Howel Williams, is printed on the reverse side of the quadrangle sheet, with accompanying geologic maps and photographs. The data given above were adapted from the Williams text.

CRATER LAKE NATL. PARK
AND VICINITY, ORE.
1:62500 C.I.= 50' 1956

CRATER LAKE

ELEVATION 6176 FEET
DEPTH ABOUT 2000 FEET

SHIP ROCK, N. MEX.   1:62500   15′   C.I. = 20 ft.   1934

When a volcano becomes inactive and ceases to erupt, the remaining magma occupying the pipe connecting the vent to the magma chamber at depth may cool and solidify to form a mass of igneous rock known as a **volcanic pipe.** If the pipe rock is more resistant to erosion than the enclosing rock units, as is often the case for a volcanic pipe surrounded by sedimentary rocks, differential erosion may lower the land surface around the pipe leaving it standing in bold relief. An eroded volcanic pipe standing out as a topographic high is commonly called a **volcanic neck.**

Ship Rock, one of the best-known landmarks in the American southwest, is a volcanic neck which towers more than 1500 feet above the surrounding landscape.

1. Two prominent ridges extend radially out from the volcanic neck. Careful study of the contour lines in the vicinity of the neck will reveal at least three other such ridges, smaller and shorter, but definitely radial to the neck. **Mark these smaller ridges with a red pencil line along the length of each ridge.**

2. These ridges are composed of igneous rock similar in composition and age to that which composes the neck. What is the proper geologic name for these **intrusive** features? Describe the origin of these ridges. (Consult your textbook and/or lab instructor.)

SHIP ROCK, N.MEX.
1:62500 15' C.I.= 20' 1934

Exercise 20: Man and His Environment

Many of the preceding exercises have been concerned with geologic processes and their effects upon the earth's surface. It is necessary for us to understand these continuing processes and their effects in order to live compatibly with them. Furthermore, it is necessary to realize that man has, in the last century, become a significant geologic agent himself, capable of altering natural processes, often without being aware of it, and often with detrimental effects to his own species.

In this exercise several maps have been selected to illustrate both how specific geologic processes affect man's endeavors, and also how human endeavor may cause large-scale alterations of the natural environment.

In an ideal society in an ideal world, man would live compatibly with geologic processes. Unfortunately, in the world as it is, man often struggles against geologic processes that don't suit his immediate needs. These "negative" geologic processes are often referred to as **geologic hazards.** Following are some of the better known geologic hazards which often disrupt human activity.

Geologic hazards

Mass movements
1. Slow mass movements: creep, solifluction, debris flows
2. Rapid mass movements: mudflows, slumps, avalanches, debris slides, rockslides and rock falls

Transportation system processes
1. Normal erosion and deposition by streams, by shoreline processes, by wind, and by ice
2. Abnormal erosion and deposition by floods, gales, and hurricanes

Surface subsidence and heaving
1. Collapse of caverns due to solution
2. Frost heaving, ice wedging, growth of pingos, etc.

Tectonic and volcanic activity
1. Slow fault movements
2. Earthquakes due to rapid fault movements
3. Volcanic eruptions
4. Tsunami ("tidal waves") due to submarine tectonic and/or volcanic activity

In addition to these natural hazards, manmade environmental hazards must be considered. The following list includes just a few of the ways in which man interferes with natural geologic processes. The student is encouraged to make further additions to the list.

Human activities which interfere with natural geologic processes:

1. Strip mining (changes topography, generally causes deterioration of local water supplies)

2. Underground mining (may cause surface subsidence and water pollution)

3. Extraction of oil and gas (may cause surface subsidence and water pollution)

4. Irrigation (may result in rapid changes in ground water levels; also may cause salt buildup in soil)

5. Dam construction (changes natural drainage, erosion, and sedimentation)

6. Paving of large areas (increases runoff, decreases input to ground water)

7. Recreational activities (often disturb or destroy natural systems)

8. Waste disposal

  a. Solid waste (landfills and garbage dumps change the natural contours of the land and pollute water)

  b. Chemical waste (industrial and agricultural wastes such as fertilizers and pesticides pollute water supplies)

  c. Gaseous waste (toxic gases and particulate matter pollute the atmosphere and may locally greatly increase weathering processes)

  d. Thermal waste (most electric power generating plants and many other industrial plants discharge heated water which is oxygen-poor and potentially damaging to life systems in the streams or lakes receiving the discharge)

Answer sheets for this exercise begin on Page A-57.

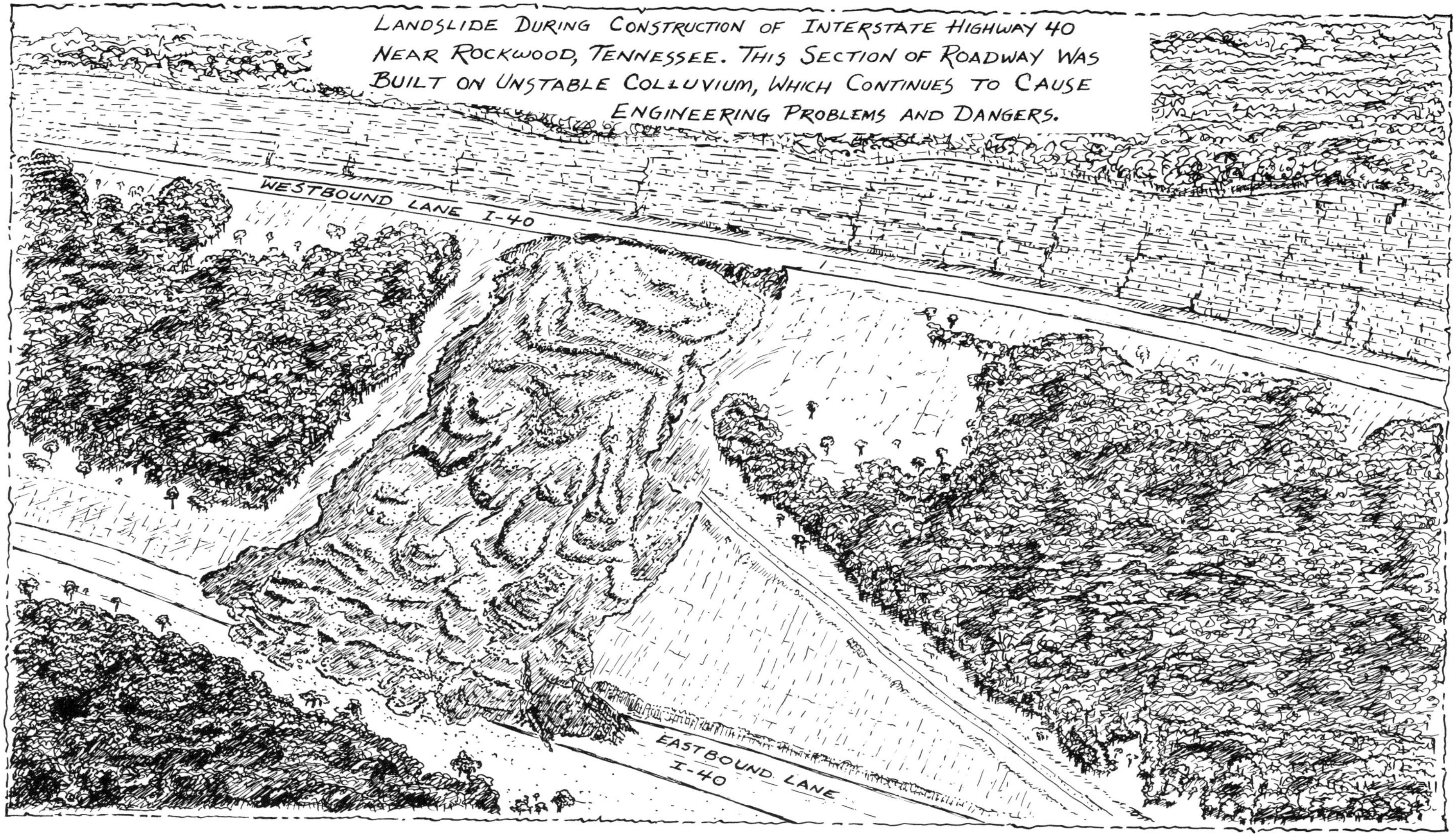

LANDSLIDE DURING CONSTRUCTION OF INTERSTATE HIGHWAY 40 NEAR ROCKWOOD, TENNESSEE. THIS SECTION OF ROADWAY WAS BUILT ON UNSTABLE COLLUVIUM, WHICH CONTINUES TO CAUSE ENGINEERING PROBLEMS AND DANGERS.

WESTBOUND LANE I-40

EASTBOUND LANE I-40

Fig. 20-1. Landslide damage to interstate highway

173

AJO, ARIZ. 1:62500 15′ C.I. = 40 ft. 1963

Ajo (pronounced *AH-hoe*; Spanish, meaning *garlic*) is a famous copper producing area. The copper minerals occur as inclusions disseminated throughout an intrusive igneous rock mass. Because the copper minerals are scattered as small blebs throughout a large rock body, rather than concentrated in veins or replacement bodies, it is necessary to mine vast quantities of rock. Therefore the mining must be done by the **open pit mining** method. Look carefully at the New Cornelia Mine; it is an immense hole in the ground.

1. How deep is the New Cornelia Mine pit? (Measure from the road on the south rim to the deepest point in the center.)

2. What is the greatest width of the New Cornelia Mine pit? (Measure along its north-south axis.)

3. That's a big hole. And most of the material taken out of this hole is not copper, but waste rock. This material must be processed by a mill, in which the copper minerals are separated from the waste. The mill shows on the map as a group of large buildings on and just north of Concentrator Hill. Notice the large **waste dump** to the southeast of Concentrator Hill. Note also the North, South, and East tailings ponds. Finely ground up waste rock which comes out of the mill after the ore minerals have been extracted is called **tailings.** Most of the rock taken out of the New Cornelia Mine pit winds up in either the waste dump or one of the tailings ponds.

Estimate the combined acreage of the North, South, and East tailings ponds. (Remember, there are 640 acres to one square mile.)

4. The copper mill (concentrator) uses immense quantities of water. But note that Ajo is a town in the middle of very arid, semi-desert country. Where do you suppose the water for the mill might come from, and what might be the long-range consequences of using such large quantities of water?

AJO, ARIZ.
1:62500 15′ C.I. = 40′ 1963

McCARTHY, ALASKA   1:250000   1° X 3° C.I. = 200 ft.   1960

(Turn to map on page 123.)

Bonanza Ridge is another copper mining area. Note the names Bonanza Peak, Porphyry Mountain, and Green Butte (the weathering of primary copper minerals commonly produces blue-green secondary copper minerals). Kennicott, aside from being the name of a cluster of buildings in this area, is also the name of one of the world's major copper mining companies.

1. Bearing in mind that this mine is located in the Wrangell Mountains of Alaska, what do you suppose would be some of the chief logistical problems associated with mining here? (Hint: Where are the roads? How is the mine supplied and how is the copper taken out of this area?)

2. What types of problems would building roads and landing strips present? (Consider the topography, the weather, and geologic processes active in this area.)

HARRIMAN, TENN.   1:24000   7½'   C.I. = 20 ft. 1968

This map segment shows many man-made changes in the natural environment: an urban area (the town of Harriman) shows in the NW corner; a super highway (Interstate 40) crosses this area; a portion of a man-made reservoir (Watts Bar Lake) is shown, along with dashed lines representing the original channel of the Clinch River; and a major electric power generating plant (the Kingston Steam Plant) is located here.

The Kingston Steam Plant is a coal-fired generating plant. This means railroads to bring in the coal, large coal stockpiling areas, and pollution problems with smoke and ash. As shown on the map, the Kingston Steam Plant has nine smokestacks; more recently two enormous smokestacks, each 1000 feet high, have been built to help disperse gaseous waste products. These stacks are equipped with "scrubbers" to help remove some of the worst pollutants. Note that the solid waste, the ash from burning the coal and the sludge from the scrubbers, must be stored in an enormous ash disposal area just north of the plant.

1. Calculate the acreage consumed thus far by the ash disposal area.

2. The electric generators are steam powered turbines. Where does the water for the steam come from and what problems might arise when the exhausted steam is condensed back to water and returned to its source?

The student should also study Figure 20-1 to see some problems encountered in the construction of Interstate 40 near Rockwood, Tennessee, immediately west of this map area.

HARRIMAN, TENN.
1:24000 7½' C.I.= 20' 1968

RIVERTON, MINN.   1:24000   7½'   C.I. = 10 ft.   1973

This is a region covered with glacial till. Most of the lakes here are of glacial origin. It is also a region of iron ore mining, of the **open pit** type. Some of the abandoned iron mines now form additional lakes.

1. List several ways in which the local inhabitants might benefit by iron ore mining in this area.

2. What are some potential environmental problems associated with iron ore mining in this area?

3. What environmental problems might arise if the town of Riverton utilized an old mine pit as its garbage dump? (Hint: What is till composed of?)

RIVERTON, MINN.
1:24000 7½' C.I. = 10' 1973

OSWEGO, KANS.   1:24000   7½'   C.I. = 10 ft.   1974

Here large areas have been disturbed by strip mining for coal. A portion of this area has been reclaimed as a wildlife management area where a habitat is provided for fish and waterfowl.

Strip mining is a type of mining in which the economic mineral resource occurs in a sedimentary bed which is overlain by beds of non-economic waste rock. These overlying beds, called **overburden,** must be stripped away by steam shovels, draglines, or bulldozers before the underlying economic bed can be mined. In the past, many strip mine operators did nothing to restore the land after stripping. Now, state and federal laws require that the land surface be restored to approximately its original contours. This requires a great deal of effort and expenditure on the part of the mine operator. It also means that the consumer will pay higher prices for the product mined, but this is necessary in order to protect and preserve the environment in which we live.

1. If we assume that the average thickness of the coal seam mined here is 3 feet and that 1800 tons of coal are produced per **acre foot,**[1] what would be the total tonnage produced from a full **section?**

2. If the profits were $5.00 per ton, what would the profit be from a complete section?

3. Obviously this area was mined prior to the enactment of proper land restoration laws. Suppose the mine operator had been required to restore this land, and the cost of restoration had been $1,000 per acre. What would the total cost of restoration be for a complete section? Do you think that this expenditure is justified or unreasonable? (Are you, as a consumer, willing to pay this cost?)

1. Ask your lab instructor about the term *acre foot* if you are not sure what it means.

OSWEGO, KANS.
1:24000 7½' C.I. = 10' 1974

FAIRBORN, OHIO   1:24000   7½'   C.I. ≈ 10 ft.   1965

This map shows an unusual land-use situation. Wright-Patterson Air Force Base has been located on the broad flat floodplain of the Mad River, an indirect tributary to the Ohio River. The normal level of the Mad River is shown by the solid blue outlines. Note that Huffman Dam, a flood control dam, has a spillway elevation of 835 feet. If this dam is closed off, a lake will form with a lake surface elevation of 835 feet; the area that would be covered by this lake is outlined by a dashed blue line.

1. What will happen to Wright-Patterson Air Force Base in the event of major flooding on the Mad River which requires that the Huffman Dam gates be closed? (If you are wondering what the Air Force will do with its planes, note that a portion of a smaller, but higher, auxillary air base shows in the SW corner of the map segment.) (Note also that there are two memorials to the Wright Brothers, one on the floodplain at about 803 feet above sea level, and the other on a bluff near the dam at about 925 feet above sea level!)

2. Can you think of any good reason why Wright-Patterson Air Force Base and a portion of the business district of Fairborn might be deliberately flooded this way? Why would a flood control dam be designed to deliberately inundate such valuable property? (Find a map of Ohio to see the answer to this question.)

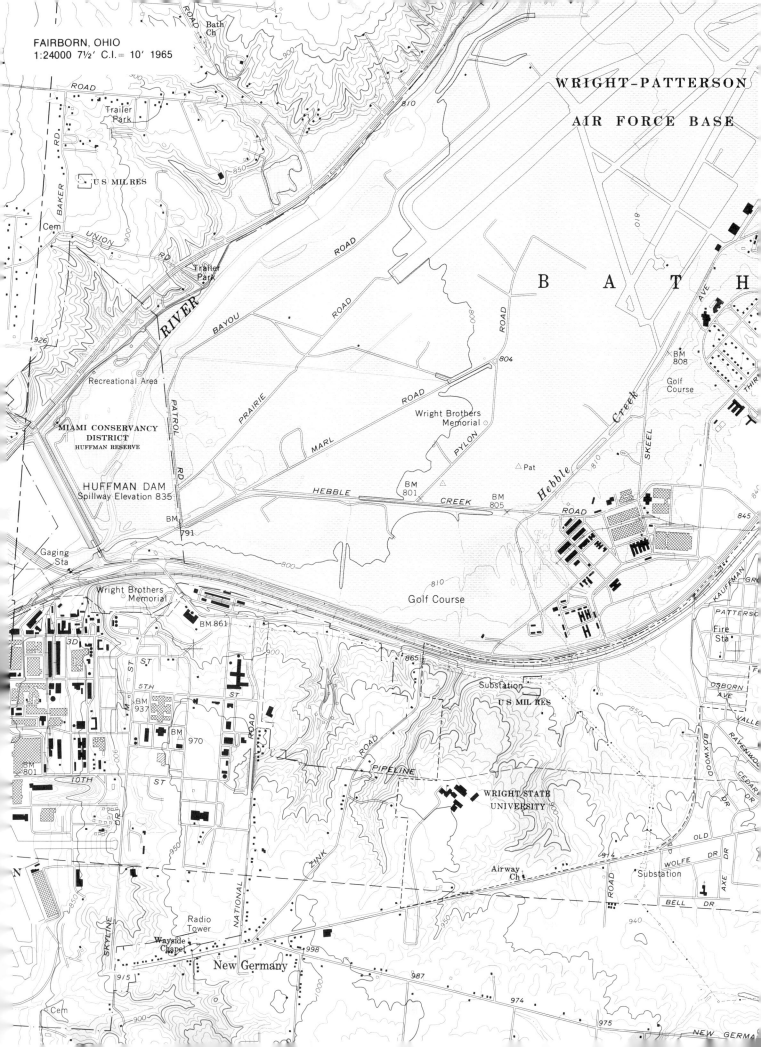

DEROUEN, LA.   1:62500   15′   C.I. = 5 ft.   1963

This map shows two geologic structures known as **salt domes,** Avery Island with a positive topographic expression, and Jefferson Island-Lake Peigneur with a largely negative topographic expression. Salt domes are masses of rock salt which move slowly upwards towards the surface, penetrating and distorting the sedimentary rock layers which surround them. There are hundreds of these salt domes in the Gulf Coast region, but only a small percentage of them reach the surface like Avery and Jefferson Islands.

Salt mining began on Avery Island in 1791 and on Jefferson Island in 1923. Jefferson Island is the home of the salt brand by that name, the brand responsible for many painted barns along Southern byways (see page 186). Avery Island is mined by the International Salt Company, producers of Morton's Salt, and the island has the further distinction of being the home of all the peppers grown for Tabasco Sauce!

In addition to salt, other economic deposits are frequently associated with salt domes: oil, gas, and sulfur. Note the oil fields associated with both Avery and Jefferson Islands. The oil fields are associated with the flanks of these two domes. Obviously since these domes have reached the surface there cannot be any oil trapped on top of them, but in many cases oil and gas do accumulate in the domed sedimentary layers over salt domes that have not reached the surface.

When sulfur is found in commercial quantities on top of salt domes, it is commonly mined by the hot water technique. Hot water is pumped down injection wells into the sulfur layer. This water melts the sulfur, which is then pumped out to the surface through extraction wells. At the surface it is allowed to cool and resolidify in tanks, to be shipped in the more convenient solid form.

1. Why do you think oil and gas might accumulate in the sedimentary layers around the flanks of these salt domes? (Remember: Oil and gas are relatively light fluids, hence they will move upward through porous and permeable rock units if allowed to do so.)

LAKE PEIGNEUR

Jefferson
Island

Gas Well

Jefferson Island

Gas Well

JEFFERSON ISLAND OIL FIELD

JEFFERSON CANAL

IBERIA PARISH

DELCAMBRE CANAL

Leleux

La Salle

BM 7

BM

Brousville

SOUTHERN

PACIFIC

MISSOURI

Bayou Petite Anse

Santiague

Deblanc Coulee

Badeaux

Norbert

SOUTHERN PACIFIC

Numa

BRANCH

ARMENCO

Cem

BM 9

BM 10

Migues

BM 10

Derouen

BM 10 Cem

Broussard Cem

Radio Tower

Nicholas

Petite

Davids

BM 6

Anse

BM 10

Ludger

Delcambre

Bob Acres

SOUTHERN

BM

PACIFIC

Poufette

BM

Hayes

Bayou

Coulee

Rynella

Marcel

BM

Cem

Emma

Cem

Landry Landing Strip

JEFFERSON CANAL

ROUFETTE CANAL

BM

Bayou Tigre

Bayou Carlin

SPOIL DEPOSIT

TIGRE LAGOON OIL AND GAS FIELD

Petite Anse

PETITE ANSE CANAL

AVERY ISLAND OIL FIELD

Anse

Tollgate

Jungle Gardens

Willow Pond

Avery Island Cem

Spence Sch

Terrell Sch

AVERY

Gas Well

Avery Island

Hayes Pond

59

57

Lake Pond

Rice Pond

ISLAND

Gas Well

Saline Woods Pond

24

Line of demarcation between salt r
and fresh marsh is not determined

Bayou

Portage

Three

Bayou

Cassmer

Bayou

PIPELINE

CANAL

SPOIL DEPOSIT

SUB PIPELINE

INTRACOASTAL

WATERWAY

AVERY

DEROUEN, LA.
1:62500 15′ C.I.= 5′ 1963

SECTION

Exercise 21: Geologic Maps and Cross Sections

In exercises 13 through 20 you learned how to recognize geologic features on topographic maps. It is now time to learn how to interpret a special type of map which shows both topography and the geology underlying the topography. The **geologic maps** in this exercise use contour lines and standard map symbols to show the topographic and cultural features; in addition to this, the rock units which crop out — or which would be exposed if there were no soil and plant cover — are indicated by the use of special colored overprints. Other geologic data such as strike and dip, fold orientations, faults, etc., are shown by **geologic map symbols** such as those presented to you in exercise 8. (It would be a good idea for the student to review exercise 8 prior to working the present exercise.)

Using geologic maps, a geologist can make intelligent speculations regarding the rocks and rock structures well below the surface of the earth; this type of scientific speculation is vital to the search for economic deposits, and critical to the resolution of many engineering problems major and minor. On a somewhat more academic plane, a geologic map can be used to interpret the **geologic history** of an area. The map shows the rocks that exist in an area and, as much as is possible, their ages; the **outcrop pattern** is also shown. From this information the geologist can interpret what this portion of the earth was like when the various rock units were formed, and he can also determine in what way the rock units have been deformed since their formation. In this manner the geologist can reconstruct the sequence of geologic events that have created and shaped that part of the world shown on the geologic map.

Sounds simple, doesn't it? Well, it is . . . in a few places. In other areas critical data may be lacking (the rock units may be poorly exposed due to dense vegetation, the ages of certain units may be unknown, the area may be so intensely faulted that the sequence of events cannot be determined with certainty, beds may be turned upside down, etc., etc.). Such data ambiguities may make the interpretation of the geologic history of an area very speculative indeed.

On the next several pages you will be presented with some block diagrams which illustrate some of the basic concepts which you must master in order to interpret geologic maps intelligently. Each block diagram is an idealized view of some segment of the earth, as if it were lifted out of the ground so you can see its sides as well as the top. **The upper surface of each block diagram is essentially a geologic map:** it shows the outcrop pattern of the rocks. **The sides of each block diagram are geologic cross sections:** these show what the rock units do at depth. The object of the game is for you to become so familiar with the relationships between the map patterns and the cross sections that you can begin to predict the cross-sectional views when given only the map view. To do this you need to think in three dimensions; you need to take outcrop patterns and other geologic data presented in the map view and project them mentally to imagine the cross-sectional view within the earth.

First of all, you need to learn the meaning of all the geologic symbols presented in Table 21-1. These are the basic symbols, and should be memorized. Many other different special purpose symbols may be seen on geologic maps, but published geologic maps will generally carry an explanation of the symbols used.

Next, study each of the block diagrams presented on the following pages, plus those in exercise 8. Notice the **outcrop patterns** which are characteristic of certain rock structures such as anticlines and synclines. Notice also the strike and dip symbols and the special symbols for folds and faults; see how they relate to the outcrop patterns. Most of these block diagrams are so highly idealized that they show only geology: no topography is shown other than a flat surface. But on the real geologic maps that you will work with, topography will be shown. The topographic relief may make the outcrop patterns more complicated than those shown on the simple block diagrams. But this topographic relief is commonly controlled in part by the underlying rock types, so the nature of the topography may be very informative. Furthermore, this topographic relief helps give you a three-dimensional view of the geology, rather than a flat, two-dimensional plan.

After studying the diagrams and symbols, and becoming familiar with such basic concepts as "the oldest beds crop out in the center of an anticline," "steep slopes make narrow outcrops," etc., you will be ready to try your hand at interpreting the five geologic maps which accompany this exercise.

Answer sheets for this exercise begin on Page A-59.

TABLE 21-1: GEOLOGIC MAP SYMBOLS

Geologic Time Units

Q — Quaternary Period

T — Tertiary Period

K — Cretaceous Period

J — Jurassic Period

Ŧ̵R — Triassic Period

Ᵽ — Permian Period

ℙ — Pennsylvanian Period

M — Mississippian Period

D — Devonian Period

S — Silurian Period

O — Ordovician Period

Є — Cambrian Period

pЄ — Precambrian

Structural Symbols

Geologic contact: this line separates two different rock units on the map when they are in normal dispositional, igneous, or metamorphic contact. Dashed = imprecisely known.

27

Strike & dip symbol: in this example the planar feature (rock layer, fault, joint, etc.) strikes North 30° West, and has a southwest dip of 27°.

Special symbol for vertically dipping feature: in this case the strike is due East-West, and the dip is, of course, 90°.

Special symbol for horizontal planar feature: strike is any and all compass directions, dip is zero.

U   D

High angle fault: could be normal or reverse; U on upthrown side, D on downdropped side. Dashed where imprecisely mapped.

Rock Type Symbols for Cross Sections

 Shale

 Limestone

 Sandstone

 Igneous Rock

 Igneous Rock

 Metamorphic Rock

Thrust fault: the "teeth" are on the hanging wall or overthrust block.

Anticline with horizontal hinge.

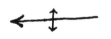

Plunging anticline: hinge inclined in direction of large arrowhead.

Syncline with horizontal hinge.

Plunging syncline: hinge inclined in direction of large arrowhead.

Fig: 21-1: Block diagrams showing some basic outcrop patterns

The block diagrams on the next few pages illustrate some of the basic outcrop patterns formed by rocks exposed at the surface of the earth. Remember: the upper portion of each block may be thought of as a geologic map in an area of zero relief (flat area). The sides of each block are geologic cross sections.

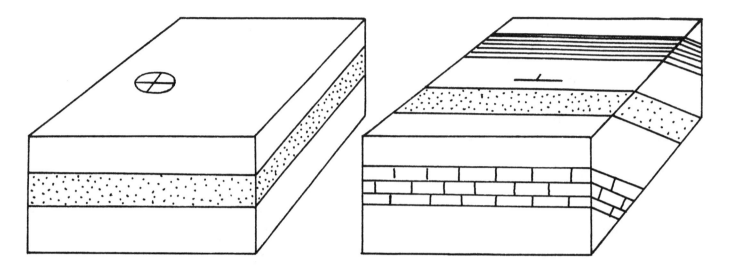

If the rock units are flat-lying, horizontal layers, only the uppermost layer will be exposed in an area of zero relief.

If the rock units are tilted layers, several units will be exposed, forming a banded outcrop pattern on the map.

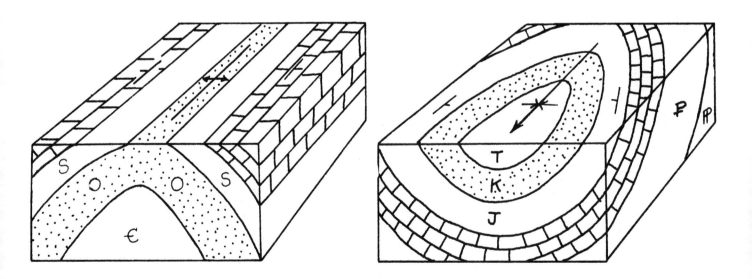

Where layered rock units have been folded into an anticline, the same units will be exposed on both sides of the fold hinge; the oldest unit will be exposed in the center of the fold. In this example, the fold hinge is horizontal, i.e., the fold is not tilted.

Where layered rock units have been folded into a syncline, the same units will be exposed on both sides of the fold hinge; the youngest unit will be exposed in the center of the fold. In this example, the fold hinge plunges toward the viewer, i.e., the fold is tilted.

Fig. 21-1: Some basic outcrop patterns, cont.

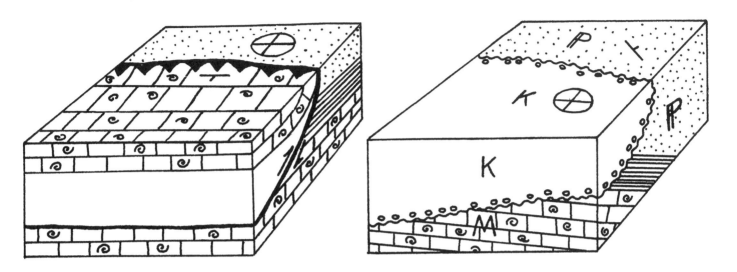

Low angle thrust fault: the map pattern shows two formations in contact which would not normally be in contact (the stratigraphic column accompanying the map would show you that these units cannot occur in sedimentary contact). Note how the thrust repeats the fossiliferous limestone layer in the cross sections.

Unconformity: the map pattern shows two formations in sedimentary contact, but separated by a surface of erosion (represented by the wavy line topped with eroded cobbles from underlying units). Although the contact might appear to be a normal stratigraphic contact at first glance, the strikes and dips, and the ages of the rock units show that it is not. This particular type of unconformity is called an **angular unconformity** because of the difference in dips above and below the surface of erosion.

Note that the map pattern shown for the unconformity is very similar to that for the thrust fault. If it were not for the special symbols we could not tell which was which. Very often there are two equally possible interpretations of a map outcrop pattern. The field geologist must have more data than just the outcrop pattern. He needs "three dimensional" data showing some of the rocks below the general surface level. These data may come from stream canyons, mines, quarries, well logs, etc. Thus he may get a peep at the "sides of the block diagram", enabling him to make the right interpretation of the geology.

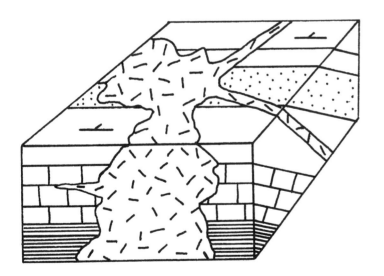

Intrusive contacts: an irregularly shaped pluton with associated dikes has intruded pre-existing rock units; the intrusive contacts cut across the other units. According to the Principle of Cross-cutting Relationships, the intrusion is the youngest of all the rock units shown in this diagram.

Fig. 21-1: Some basic outcrop patterns, cont.

Repetition and Omission of Beds by High Angle Faults

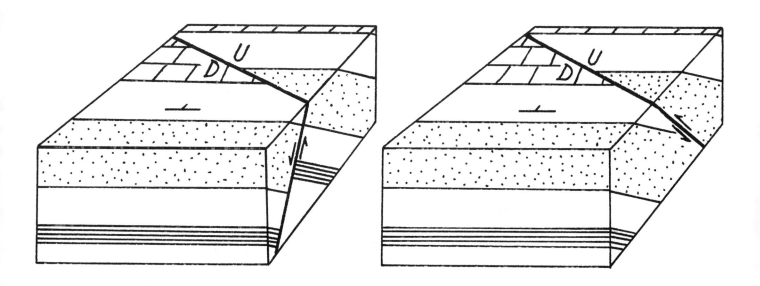

The two block diagrams above show **repetition of beds** and identical outcrop patterns. On the left, repetition of beds is caused by a **normal fault dipping in the opposite direction from the bedding dip.** To the right, repetition of beds is caused by a **reverse fault dipping in the same direction as the bedding dip.** Note that based on the outcrop pattern alone, the faults might be thought to be left-lateral strike slip. Obviously you must have more than just the map pattern to successfully interpret the geology and draw in the sides of these blocks.

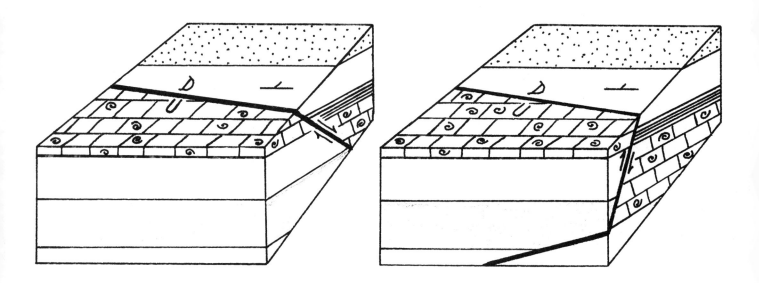

The two block diagrams above show **omission of beds:** the thin shale bed does not crop out at the surface of either block. On the left, it has been dropped down out of sight in the hanging wall of the normal fault; on the right, it has been covered up by the hanging wall of the reverse fault. Again the map patterns are identical, and more data is required than just the outcrop pattern in order to interpret the geology.

Fig. 21-2: The Rules of the "Vees" for Layered Rocks: Outcrop patterns formed where rock layers cross stream valleys.

Block Diagrams   **Outcrop Pattern on Map**

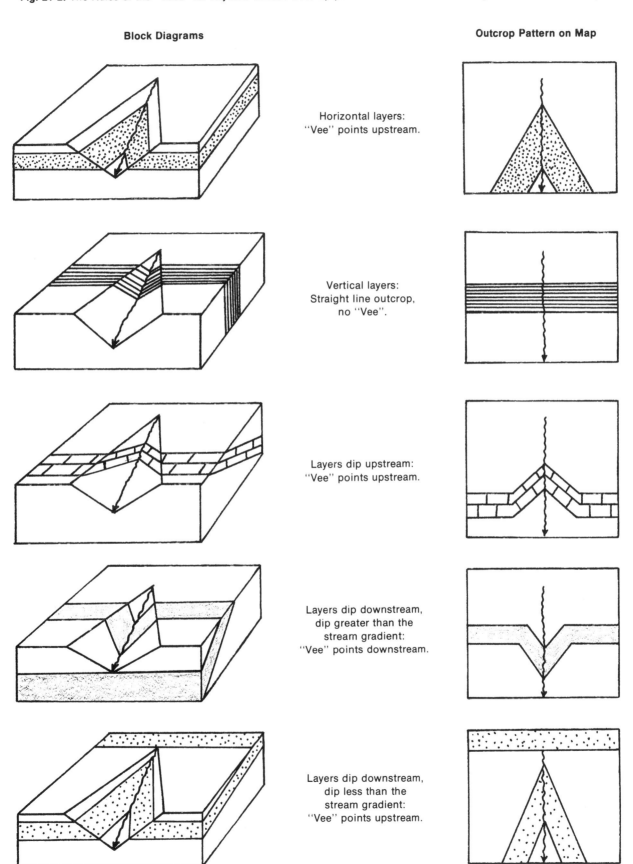

Horizontal layers:
"Vee" points upstream.

Vertical layers:
Straight line outcrop,
no "Vee".

Layers dip upstream:
"Vee" points upstream.

Layers dip downstream,
dip greater than the
stream gradient:
"Vee" points downstream.

Layers dip downstream,
dip less than the
stream gradient:
"Vee" points upstream.

FLAGSTAFF PEAK, UTAH   1:24000   7½′   C.I. = 40 ft.   1979 GEOLOGIC QUAD.

The different colors on this map represent different rock layers called **formations.** To qualify as a formation, a rock unit must have distinctive characteristics — such as composition, bedding, color, fossil content, etc. — which enable the field geologist to recognize and distinguish it from other rock units in the area. (Another technical requirement of a formation is that it be mappable at a scale of 1:25000. That is to say, a rock body too small or too thin to be represented on a map at a scale of 1:25000 cannot be called a formation. Why 1:25000? Most of the civilized world uses the metric system, not the English system, and multiples of 12 aren't very handy in the metric system. Geologists are part of a worldwide scientific community and the rules are written with the worldwide community in mind.)

There are seven formations cropping out in this area. In addition to the use of distinctive colors, each is identified on the map by a code, such as "Ksp." The upper case letter indicates the age of the formation: "K" stands for Cretaceous Period. The lower case letter or letters that follow stand for the name of the formation: "sp" is the abbreviation for "Star Point Sandstone," a sandstone formation first described and defined near Star Point, a prominent landmark in the Gentry Mountains near Price, Utah. Accompanying most geologic maps is a **stratigraphic column** consisting of colored boxes; beside each box is the full formation name, plus, in many cases, a brief description of the rock unit. See the stratigraphic column below.

The Flagstaff Peak geologic quadrangle is a splendid example of "layer cake geology." Note that the contact lines separating the various formations tend to parallel the topographic contour lines. This shows you that the contact or surface between two adjacent formations is a nearly horizontal plane. This should make drawing a **geologic cross section** pretty easy in this map area.

In the answer sheets for this exercise, you will find a topographic profile drawn along line A-A'. The profile has no vertical exaggeration, and shows you the hills and valleys in true proportions. Convert this profile into a geologic cross section, following the instructions below the profile line. Remember, the map is like the top view of a block diagram (but one with topography). What you want to do is to project the map data "underground" to create the cross-sectional view, which is like the side of a block diagram.

Stratigraphic Column for Flagstaff Peak Geologic Quad

| Qal | Quaternary alluvium: gravel, sand, silt, & clay deposited in stream valleys. |

| TKn | North Horn Formation (Paleocene and Upper Cretaceous): variegated shale with interbedded limestone and conglomerate. Thickness about 1300 ft. |

| Kpr | Price River Formation (Upper Cretaceous): gray- to light-gray sandstone, coarse-grained to conglomeratic, with some shale. Thickness about 295 ft. |

| Kc | Castlegate Sandstone (Upper Cretaceous): coarse-grained, massive-bedded, cliff-forming, weathers grayish-orange. Thickness 100 to 195 ft. |

| Kb | Blackhawk Formation (Upper Cretaceous): grayish-orange to medium-gray, medium- to fine-grained sandstone with some coal. Thickness about 770 ft. |

| Ksp | Star Point Sandstone (Upper Cretaceous): cliff-forming, gray to grayish-orange sandstone, siltstone, and shale. Thickness 245 to 310 ft. |

| Kmm | Mancos Shale, Masuk Member (Upper Cretaceous): medium-to dark-gray shale, weathers yellowish gray. Thickness about 850 ft. |

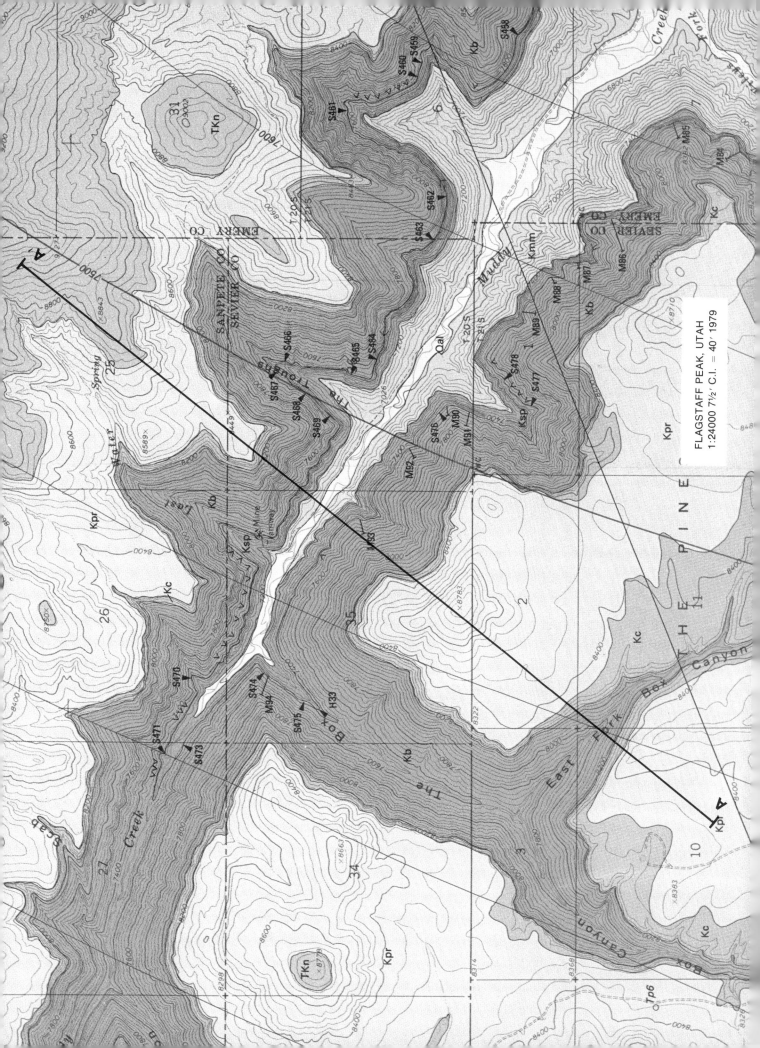

FLAGSTAFF PEAK, UTAH
1:24000 7½' C.I. = 40' 1979

WETTERHORN PEAK, COLO. 1:24000 7½' C.I. = 40 ft. 1972 GEOLOGIC QUAD.

Three Tertiary rock units are exposed in this map area: the San Juan Formation (Tsj), consisting of gray to red, massive bedded tuffs, volcanic breccias, and andesite flow rocks; the Potosi Formation (Tp), consisting of gray to brown ignimbrites of latitic composition; and several bodies of porphyritic quartz latite (Tql). ("Latite" is a compositional term similar to rhyolite, but latitic rocks fall somewhere between rhyolite and andesite on the igneous rock classification chart.) Other deposits shown on this map include several types of unconsolidated Quaternary deposits: talus (Qt), stream alluvium (Qal), landslide deposits (Qs), and alluvial-cone deposits (Qac).

1. Examine the contacts of the porphyritic quartz latite bodies (Tql) where they contact the San Juan and Potosi formations. What type of igneous unit is the quartz latite: extrusive or intrusive? Explain your reasoning.

2. The long, narrow bodies of Tql which radiate out from the two main quartz latite bodies are known as___?

3. Note how the contact between the San Juan and Potosi formations in the NW portion of the map tends to parallel the contour lines. This indicates that the contact between the two units approximates a horizontal plane. (Be sure you understand this; ask your lab instructor if you do not.) Are these two units extrusive or intrusive? Explain your reasoning.

4. Applying the Principle of Cross-cutting Relationships, the Principle of Superposition, and any other pertinent data (such as the alteration of Tsj rocks between the main Tql bodies), determine the **relative ages** of the three Tertiary igneous units. List these units in a vertical column, oldest on the bottom, youngest on top.

5. The main Tql body seems to be unusually rectangular, being especially linear along its east and south sides. Place a ruler along the east side of this body and note how a long dike follows this same trend to the north; now place the ruler parallel to the south side and note the dikes extending in both directions along this trend. What type of **structure** might control these dike orientations and even influence the emplacement of the latite intrusion?

6. Note the fault cutting across the San Juan and Potosi formations in the NE corner of the map. What is the dip of the fault plane: low angle, high angle, or nearly vertical? How can you tell?

7. In the SE portion of the map, at the letter "C," there is an almost circular deposit of talus, partially surrounded by a semicircle of cliffs. Examine the topography shown by the contour lines surrounding this deposit and the similar deposit just to the NW. What do you think has happened here in the not-too-distant geologic past? (You may wish to refer back to exercise 16 to answer this question.)

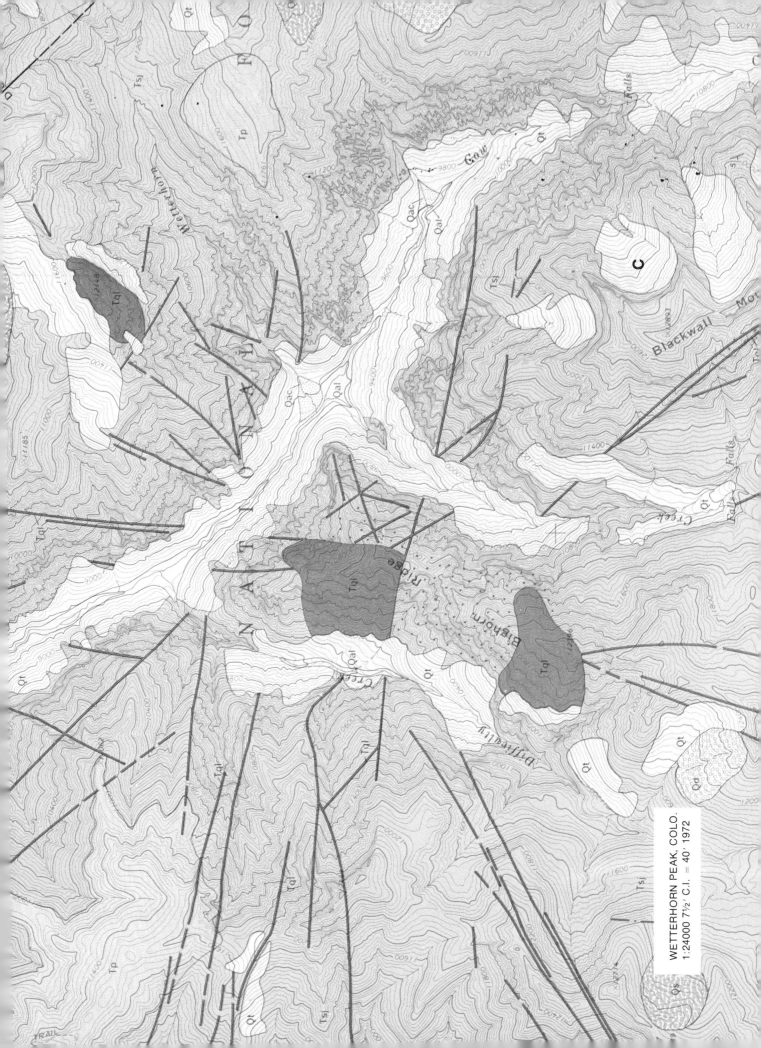

WETTERHORN PEAK, COLO.
1:24000 7½' C.I. = 40' 1972

MIFFLINTOWN, PA. 1:24000 15′ C.I. = 20 ft. 1962 GEOLOGIC QUAD.

This part of Pennsylvania belongs to the Valley and Ridge Province of the Appalachian structural belt. The Valley and Ridge physiographic province is characterized, as the name implies, by parallel ridges with intervening valleys. The linear nature of the valley and ridge topography is due to the fact that the eroded edges of a number of different sedimentary layers are exposed at the surface in this region. The ridges are held up by resistant rock units, such as the Tuscarora Formation (St), a tough, quartz-cemented sandstone. The valleys are underlain by formations which are less resistant to erosion, such as the Reedsville Formation (Or), a shaly unit. Thus the present topographic relief is largely the result of streams eroding down into the non-resistant units to form valleys, leaving the hard-to-erode units standing high in ridges.

1. Are the rock layers in this area flat-lying, simply tilted, or folded? Support your answer with at least two lines of reasoning or evidence.

2. Closely examine the outcrop pattern along cross-section line B-B′. Does this line cross any folds? If so, what kind of folds (anticlines or synclines) and how many?

3. In red pencil, draw in on your map the proper symbol for each anticline and/or syncline crossed by the section line. The fold symbols should be located in the center or hinge zone of the folds, and should be oriented parallel to the folds.

4. Construct a geologic cross section along line B-B′, using the topographic profile provided for you in the answer sheets. This cross section will be considerably more complicated than the "layer cake geology" cross section. One suggestion will help you: assume that the stratigraphic thickness of each formation remains constant throughout the cross-section area, neither thickening nor thinning noticeably.

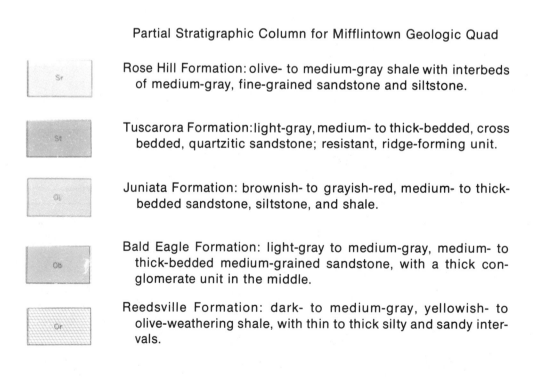

Partial Stratigraphic Column for Mifflintown Geologic Quad

Rose Hill Formation: olive- to medium-gray shale with interbeds of medium-gray, fine-grained sandstone and siltstone.

Tuscarora Formation: light-gray, medium- to thick-bedded, cross bedded, quartzitic sandstone; resistant, ridge-forming unit.

Juniata Formation: brownish- to grayish-red, medium- to thick-bedded sandstone, siltstone, and shale.

Bald Eagle Formation: light-gray to medium-gray, medium- to thick-bedded medium-grained sandstone, with a thick conglomerate unit in the middle.

Reedsville Formation: dark- to medium-gray, yellowish- to olive-weathering shale, with thin to thick silty and sandy intervals.

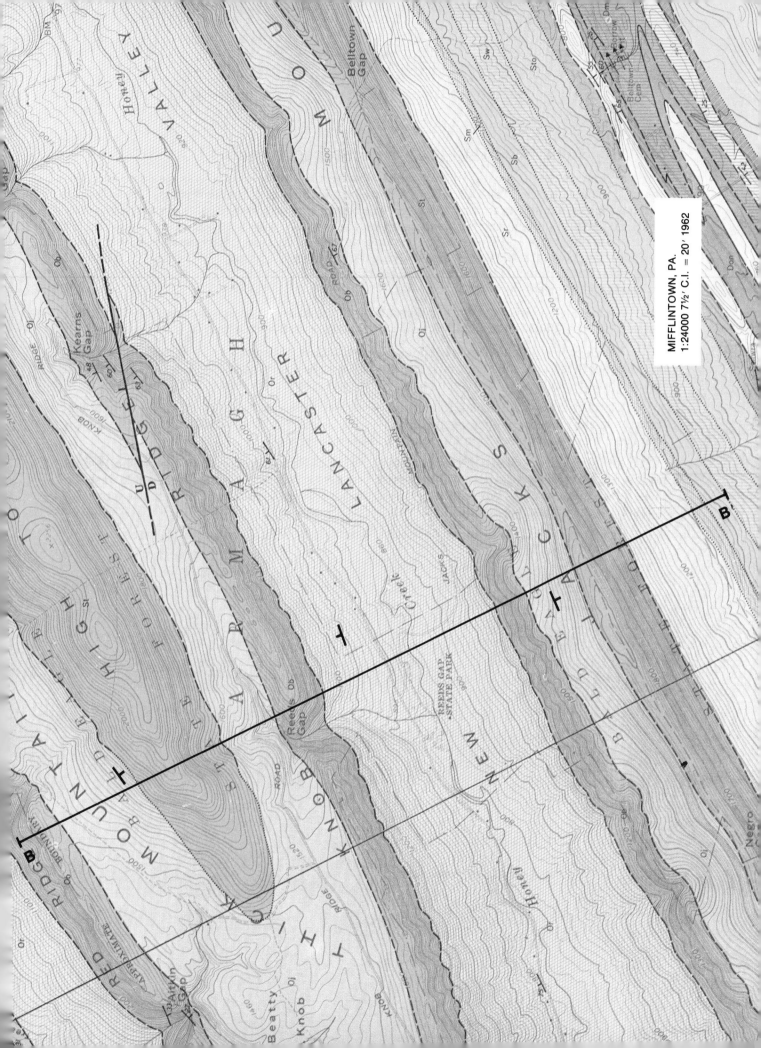

MIFFLINTOWN, PA.
1:24000 7½′ C.I. = 20′ 1962

DANFORTH, MAINE   1:62500   15′   C.I. = 20 ft.   1963 GEOLOGIC QUAD.

The Danforth area is located in an igneous-metamorphic terrane which was glaciated by continental ice sheets during Pleistocene time (approximately the last 2.5 million years). Today glacial deposits blanket almost the entire quadrangle. The bedrock geology must be interpreted from widely scattered, small outcrops which are indicated on the map by the small patches of darker shaded colors. Note how strike and dip, cleavage attitude, and other structural information is almost invariably associated with these bedrock areas. The rest of the map is an interpretation and projection of these data. This explains why many of the contacts and fault traces are shown as dashed lines, indicating that they are imprecisely known.

1. What is the nature of the large fold structure in this area: is it an anticline or a syncline? How can you tell from the outcrop pattern? How can you tell from the structural map symbols?

2. The Devonian granite (Dg) is younger than the Ordovician and Silurian metasediments. Suppose none of these rock units had labels showing their ages. There are several ways you could prove that the granite is younger than the metasediments even without the labels. Explain two ways.

3. A large fault cuts across the southwestern portion of the map area. What is the nature of this fault: strike-slip, dip-slip, or oblique-slip? Explain your reasoning. (A review of the block diagrams in exercise 8 may help you envision the solution to this problem.)

4. Just south of the village of Danforth find Horseback Road, which follows the crest of a long, narrow ridge of stratified glacial drift. "Horseback" is a New England colloquialism used to describe these long, sinuous ridges. What is the geologic name for such a feature? (Refer back to exercise 17 if you don't remember this curious feature.)

5. Name two other features shown on the map which is indicative of continental glaciation.

Zone of contact metamorphism

Glacial striae

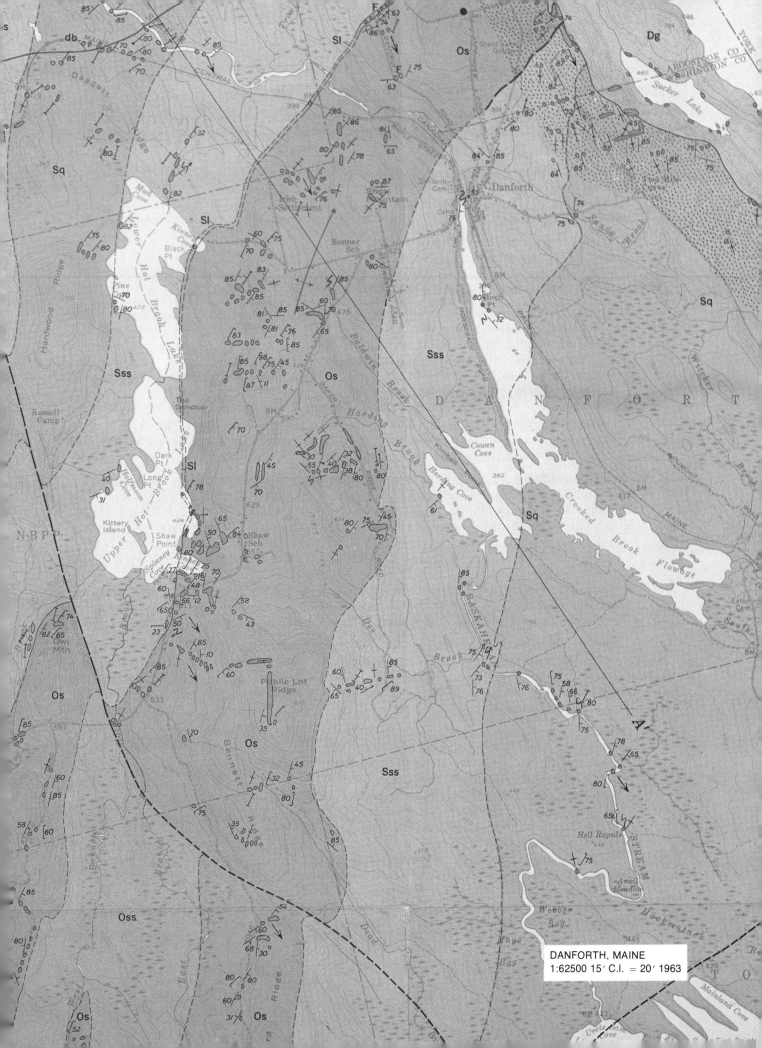

DANFORTH, MAINE
1:62500 15′ C.I. = 20′ 1963

BRISBIN, MONT. 1:24000 7½' C.I. = 40 ft. 1964 GEOLOGIC QUAD.

The Brisbin quadrangle is underlain by a very wide range of rock types, deposited during a very long time span. These rocks have been deformed in several ways, at different times. How much of the geologic history of the area can you reconstruct from the map?

1. The oldest rock unit exposed in this map area:
    a. What is the age of the oldest rock unit?
    b. What type of rock is it: sedimentary, intrusive, extrusive, or metamorphic?
    c. This unit is exposed in the middle of what type of structure: a syncline or an anticline?

2. The youngest geologic unit shown on the map:
    a. What is the age of the youngest unit?
    b. Is the youngest deposit a lithified rock unit, or is it an unconsolidated sediment?
    c. Where is this deposit found and how did it get there?

3. Cropping out in this area are formations which represent all of the geologic periods from Cambrian to Cretaceous, except for three.
    a. Name the three periods not represented by rock units in this area, listing them in a vertical column with the oldest on the bottom, youngest on top.
    b. Suggest some reason for there not being any rocks of these periods deposited or preserved here.
    c. The letter "U" marks the contact between the Bighorn Dolomite (Ordovician) and the Jefferson Dolomite (Devonian). (This is one of the places where one of the periods is not represented by a rock unit.) When a rock unit of a certain age is deposited directly onto much older rocks, the surface separating the two units is known as___?

4. Two types of faults are shown in this area: thrust faults and high angle normal faults.
    a. What type of fault is the east-west trending fault marked "T"?
    b. South Dry Creek Fault is an example of which type of fault?
    c. Which of the faults is the older, the thrust fault or the high angle normal fault? Explain your reasoning.
    d. Using the "Rules of Vees" you can tell which way the thrust fault plane is dipping. Is the dip of the fault north, south, east or west?
    e. Knowing the general direction of the thrust fault dip, you should be able to say which way the hanging wall was pushed. Which direction did the hanging wall or overthrust block move: northward, southward, eastward, or westward?

5. Using all the information determined in the preceding problems, plus any other information available to you on the map, reconstruct the geologic history of this area. A series of geologic events is listed below, but not in the order in which these events occurred. List these events in order, the first (oldest) event at the bottom of the list, then the next event, etc., until you get to the latest (youngest) event at the top of the list.

> Cambrian and Ordovician strata deposited.
> High angle normal faulting occurs.
> Quaternary alluvium and colluvium (Qac) deposited on pediment surface.
> Jurassic and Cretaceous formations deposited.
> Folding and uplift occurs; Silurian units eroded away at this time (if ever deposited).
> Granitic intrusion forms part of the Precambrian "basement" complex.
> Permian and Triassic rocks deposited (?)
> Silurian rocks deposited (?)
> Thrust faulting occurs during Laramide orogeny.
> Qya deposited by modern streams.
> Renewed uplift and more folding; Permian and Triassic rocks eroded away at this time (if ever deposited).
> Devonian, Mississippian, and Pennsylvanian formations deposited.

BRISBIN, MONT.
1:24000 7½' C.I. = 40' 1969

Answer Sheets

EXERCISE 1:
MINERAL PROPERTIES

NAME _____ LAB SEC.

1. Listed below are 8 different materials. Indicate which are minerals and which are not minerals. For those which are not minerals, explain why not.

Material	Mineral or Non-Mineral	Explanation
Synthetic Quartz	_____	_____
Wood	_____	_____
Ice Cubes	_____	_____
Table Salt	_____	_____
Petroleum	_____	_____
Brass	_____	_____
Window Glass	_____	_____
Pyrite (Fool's gold)	_____	_____

2. Examine the mineral specimens furnished by your instructor. Describe the crystal form of each mineral below.

Mineral Name or Sample Number	Crystal Form
_____	_____
_____	_____
_____	_____
_____	_____
_____	_____
_____	_____
_____	_____

3. Examine the mineral specimens furnished by your instructor. Describe the luster of each mineral below. If the mineral has a nonmetallic luster, list the type nonmetallic luster as shown in Table 1-1.

Mineral Name or Sample Number   **Luster**

_____   _____

_____   _____

_____   _____

_____   _____

_____   _____

_____   _____

_____   _____

_____   _____

_____   _____

4. Examine the mineral specimens furnished by your instructor. For each of the minerals, determine the fracture and/or cleavage type. If the sample has cleavage, list the number of cleavage directions. For those with two or three directions of cleavage, estimate the angle between the directions.

Mineral Name or Sample Number	Fracture and/or Cleavage	No. of Cleavage Directions	Cleavage Angles
_____	_____	_____	_____
_____	_____	_____	_____
_____	_____	_____	_____
_____	_____	_____	_____
_____	_____	_____	_____
_____	_____	_____	_____
_____	_____	_____	_____
_____	_____	_____	_____

EXERCISE 1, Continued

5. Examine the mineral specimens furnished by your instructor. For each of the minerals determine the hardness using common materials as listed in Table 1-2. If a set of standard hardness minerals is available, determine the hardness more accurately.

Mineral Name or Sample Number	Hardness using common materials <2½, 2½-3½, 3½-5½, >5½	Hardness using standard set
_____	_____	_____
_____	_____	_____
_____	_____	_____
_____	_____	_____
_____	_____	_____
_____	_____	_____
_____	_____	_____
_____	_____	_____
_____	_____	_____
_____	_____	_____

EXERCISE 2: MINERAL IDENTIFICATION

NAME _____ LAB SEC. _____

Sample Number	Luster	Streak	Hardness	Color	Cleavage/Other Properties	Mineral Name and Composition

MINERAL IDENTIFICATION

Sample Number	Luster	Streak	Hardness	Color	Cleavage/Other Properties	Mineral Name And Composition

A-6

MINERAL IDENTIFICATION

NAME _____ **LAB SEC.** _____

Sample Number	Luster	Streak	Hardness	Color	Cleavage/Other Properties	Mineral Name And Composition

MINERAL IDENTIFICATION

Sample Number	Luster	Streak	Hardness	Color	Cleavage/Other Properties	Mineral Name and Composition

A-8

MINERAL IDENTIFICATION

NAME _____ **LAB SEC.** _____

Sample Number	Luster	Streak	Hardness	Color	Cleavage/Other Properties	Mineral Name And Composition

MINERAL IDENTIFICATION

Sample Number	Luster	Streak	Hardness	Color	Cleavage/Other Properties	Mineral Name And Composition

EXERCISE 3:
INTRODUCTION TO ROCKS

    This exercise will introduce the use of texture and compositional differences in the study of rocks. Examine the rock specimens furnished by your instructor. Keep in mind that most mineral components in a rock have formed in a physically restricted environment, therefore well-defined crystal form may not be present. The other physical properties should not change.

Specimen No. 1

    A. Before you is a rock specimen. How could you best describe it? Using your own words, briefly describe the rock:

_____

_____

_____

Did you mention the shape and size relationships of the mineral grains? Did you mention the fact that it is composed of different minerals?

B. The jar accompanying the previous rock contains fragments obtained by crushing it. Spread the contents of the jar on a piece of paper. Analyze the compositional materials more closely. Separate the crushed materials into piles with similar characteristics.

What criteria did you use to separate these materials?

_____

_____

Examine the material in each pile very carefully. List the characteristics of the material in each pile.

Pile # 1 _____

Pile # 2 _____

Pile # 3 _____

Pile # 4 _____

Each pile should represent a different mineral. Using the Mineral Identification Tables from Exercise 2, identify each mineral.

Mineral # 1   Mineral # 2   Mineral # 3   Mineral # 4

_____   _____   _____   _____

    You will use the same process of mineral identification in future exercises without physically separating the minerals.

    C. Based on the texture described in Part A and the composition derived in Part B, is the original rock igneous, sedimentary, or metamorphic?

_____

Specimen No. 2

A. Before you is a rock specimen. How could you best describe it? Using your own words, briefly describe the rock:

_____

_____

_____

Did you mention the shape and size relationships of the mineral grains? Did you mention the fact that it is composed of different minerals?

B. The jar accompanying the previous rock contains fragments obtained by crushing it. Spread the contents of the jar on a piece of paper. Analyze the compositional materials more closely. Separate the crushed materials into piles with similar characteristics.

What criteria did you use to separate these materials?

_____

_____

Examine the material in each pile very carefully. List the characteristics of the material in each pile.

Pile # 1 _____

Pile # 2 _____

Pile # 3 _____

Pile # 4 _____

Each pile should represent a different mineral. Using the Mineral Identification Tables from Exercise 2, identify each mineral.

Mineral # 1	Mineral # 2	Mineral # 3	Mineral # 4
_____	_____	_____	_____

You will use the same process of mineral identification in future exercises without physically separating the minerals.

C. Based on the texture described in Part A and the composition derived in Part B, is the original rock igneous, sedimentary, or metamorphic?

_____

NAME _____   LAB SEC. _____

Specimen No. 3

A. Before you is a rock specimen. How could you best describe it? Using your own words, briefly describe the rock.

_____

_____

_____

Did you mention the shape and size relationships of the mineral grains? Did you mention the fact that it is composed of different minerals?

B. The jar accompanying the previous rock contains fragments obtained by crushing it. Spread the contents of the jar on a piece of paper. Analyze the compositional materials more closely. Separate the crushed materials into piles with similar characteristics.

What criteria did you use to separate these materials?

_____

_____

Examine the material in each pile very carefully. List the characteristics of the material in each pile.

Pile # 1 _____

Pile # 2 _____

Pile # 3 _____

Pile # 4 _____

Each pile should represent a different mineral. Using the Mineral Identification Tables from Exercise 2, identify each mineral.

Mineral # 1   Mineral #2   Mineral # 3   Mineral # 4

_____   _____   _____   _____

You will use the same process of mineral identification in future exercises without physically separating the minerals.

C. Based on the texture described in Part A and the composition derived in Part B, is the original rock igneous, sedimentary or metamorphic?

_____

EXERCISE 4: IGNEOUS ROCKS

NAME _____   LAB SEC. _____

| Sample Number | Texture | Color | Composition | | | Rock Name |
			Minerals	Approx. Percentage		

IGNEOUS ROCK IDENTIFICATION

| Sample Number | Texture | Color | Composition | | | Rock Name |
			Minerals	Approx. Percentage		

EXERCISE 5: SEDIMENTARY ROCKS

NAME _____ LAB SEC. _____

Sample Number	Texture	Particle Size	Composition	Other Features	Rock Name

A-17

SEDIMENTARY ROCK IDENTIFICATION

Sample Number	Texture	Particle Size	Composition	Other Features	Rock Name

EXERCISE 6: METAMORPHIC ROCKS

NAME _____

LAB SEC. _____

Sample Number	Texture	Particle Size	Composition	Other Features	Rock Name	Possible Original Rock

METAMORPHIC ROCK IDENTIFICATION

Sample Number	Texture	Particle Size	Composition	Other Features	Rock Name	Possible Original Rock

NAME _____ LAB SEC. _____

Study the specimens of common ore minerals and other natural resources which are on display in the laboratory. For each specimen, record the composition and list its most important uses. (See Tables 2-2 and 2-3, and consult your textbook and lab instructor.)

No.	Sample	Composition	Important Uses
1	Galena		
2	Chalcopyrite		
3	Magnetite		
4	Pyrite		
5	Sphalerite		
6	Hematite		
7	Fluorite		
8	Barite		
9	Muscovite		
10	Limestone		
11	Clay		
12	Bauxite		
13	Silica sand		
14	Gypsum		
15	Graphite		
16	Coal		
17	Oil shale		
18	Crude oil		

1. Measure and record the strike and dip of two planar surfaces selected by your lab instructor.

2. In the rectangle to the right, record the same two strike and dip measurements, but this time use map symbols instead of writing them out. Assume the rectangle is a map, and north is towards the "top" of the page, as is the case with most maps.

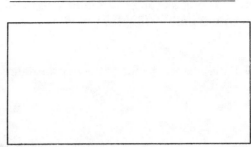

3. Classify the faults shown in the block diagrams below, as normal, reverse, thrust, left-lateral strike-slip, or right-lateral strike-slip faults.

4. Indicate with half arrows the direction of relative movement along the fault planes; indicate, where appropriate, the upthrown and downthrown blocks with the letters *U* and *D* respectively; label, where appropriate, the hanging wall and footwall sides of the faults.

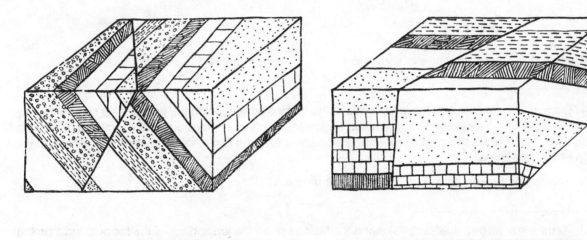

Type Of Fault   Type Of Fault

5. You may have noticed while answering questions 3 & 4, that the displacements shown in the two block diagrams can be produced by more than one type of fault movement. Actually there are two possible answers for the "type of fault" that produced each of the displacements shown, both answers being equally correct as far as the relative displacements are concerned. (To determine the actual type of movement that occurred would require more data, which the geologist might find in the field, but which cannot be represented on these block diagrams.)

What are the other fault types that could account for the displacements shown?

6. What type of fold structure is shown in the block diagram below?

_____

7. Draw in the fold hinge on the top surface of the block diagram, using the proper map symbol for this type fold. Place some strike and dip symbols on each side of the fold hinge to indicate the general strike and dip of the fold limbs.

8. This fold is broken by a fault, and in this case there is only one possible correct interpretation of the type of fault. What type of fault is it?

_____

9. Label the fault with all the appropriate symbols. Also label the hanging wall and footwall sides of the fault.

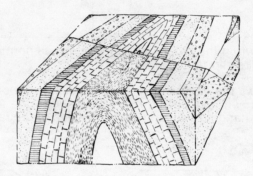

10. What type of fold structure is shown in the block diagram below?

_____

11. Draw in the fold hinge on the top surface of the block diagram, using the proper map symbol for this fold; note that it will be necessary to draw the fold symbol in two offset portions due to the fault that offsets the fold structure. Place some strike and dip symbols on each side of the fold hinge to indicate the general strike and dip of the fold limbs.

12. This fold is broken by a fault. What type of fault is it?

_____

13. Place shear arrows (half-arrows) along the fault trace on the top surface of the block to indicate the direction of relative displacement along the fault.

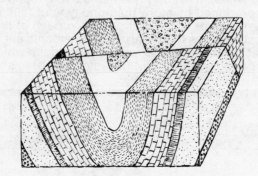

NAME   LAB SEC.

MENAN BUTTES, IDAHO, STEREOPAIR AND TOPOGRAPHIC MAP

1. DO ON STEREOPAIR. LAB INSTRUCTOR MAY INSPECT.

2. Which slope is steeper, the northeast or the western? _____

Describe the spacing of the contour lines used to show this steeper slope on the Menan Buttes topographic map.

_____

_____

COOKEVILLE EAST, TENN. STEREOPAIR AND TOPOGRAPHIC MAP

1. DO ON STEREOPAIR AND MAP. LAB INSTRUCTOR MAY INSPECT.

BRIGHT ANGEL, ARIZ. STEREOPAIR AND TOPOGRAPHIC MAP

1. Write a brief description of the topography shown in the aerial photos.

_____

_____

_____

2. Which shows the topography in more **detail** (larger scale), the stereopair or the topographic map?

_____

3. DO ON TOPOGRAPHIC MAP. LAB INSTRUCTOR MAY INSPECT.

EXERCISE 10:
INTRODUCTION TO MAP READING

NAME _____

LAB SEC.

Equipment: Triangle, scale, **sharp** 3-H pencil, scratch paper, protractor.

GENERAL PROBLEMS

1. Convert the following representative fractions to verbal statements in which ground distance equals one inch on map.

    a. 1:12000   1″ = _____ feet.

    b. 1:126720   1″ = _____ miles.

    c. 1:500000   1″ = _____ miles.

2. Express as a ratio each of the following statements.

    a. 1 centimeter represents 600 centimeters _____

    b. 1 centimeter represents 3 kilometers _____

    c. 1 meter represents 600 kilometers _____

    d. 1 inch represents 3 miles _____

PHYSIOGRAPHIC MAP OF TENNESSEE

(Turn to map on page 57.)

1. What is the scale of this map? _____

2. Measure the distance from Nashville to Knoxville. _____

3. Compare the probable accuracy of measurements on this map with the Cookeville East Quadrangle. Which is most accurate?

_____

Look at the areas covered by each map and the different methods of portraying map features.

COOKEVILLE EAST, TENN.

(Turn to map on page 43.)

This map represents a **portion** of the Cookeville East 7½ minute quadrangle. The latitude and longitude of the "cross" near the head of Rockwell Hollow is 36° 10′ 00″ N and 85° 25′ 00″ W, respectively. The "cross" just east of Whitson Chapel School has the coordinates of latitude 36° 10′ 00″ N and longitude 85° 27′ 30″ W. Assume the bottom (south) edge of the map page is latitude 36° 07′ 30″ N.

1. What is the latitude and longitude of Quarles Cemetery on this map?

_____

2. How many degrees, minutes, and seconds of latitude are covered by this map? Longitude?

_____   _____

3. What is the scale of this map? _____

4. How far is it, in miles, from the Cookeville filtration plant to the Mt. Pleasant School on this map?

_____

5. Estimate the distance by road from the filtration plant to the school._____

6. How many square miles does this map cover?_____

7. Figure the width in miles of the **entire 7½ minute quadrangle.** (Remember: The distance between the two crosses is 2½ minutes.)

_____

8. Divide this portion of the Cookeville East Quadrangle into 9 equal rectangles. Further subdivide the large rectangle containing the Mt. Pleasant School into nine smaller rectangles. What are the **rectangle numbers** for the School location?

_____

What are the **rectangle letters** for this location?

_____

9. Locate Poplar Grove Church by the same method.

Rectangle numbers: _____

Rectangle letters: _____

10. Complete the following table using this portion of the Cookeville East Quadrangle and the standard map symbols on page 55.

Feature	Map Symbol	Name An Example Found On This Map
Church	_____	_____
School	_____	_____
Cemetery	_____	_____
Major man-made structure	_____	_____
Water feature other than a stream	_____	_____

EXERCISE 10, Continued   NAME   LAB SEC.

BRIGHT ANGEL, ARIZ.

(Turn to map on page 45.)

1. Why are there no section lines for most of the map? (Why do the section lines stop at the canyon rim?) _____

_____

2. What is the scale of the map?

_____

3. What is the distance in miles between Isis Temple and Zoroaster Temple?

_____

4. How many inches apart would these features be on a map with the same scale as the Cookeville East Quadrangle?

_____

5. What is the significance of the green dot pattern in the northeast position of this map?

_____

6. When was this map published? _____

7. How many square miles does this map represent? _____

8. Give the location of Pipe Spring using bearing and distance from the visitor center.

_____

9. Using a blue pencil, show the map symbol for a spring.

_____

A-29

STRASBURG, VIRGINIA

(Turn to map on page 58.)

1. What is the scale of this map? _____

2. What is the distance in miles between the northeasternmost ford on the North Fork of the Shenandoah River in the northwest part of the map and Shawl Gap in the eastern part of the map?

_____

3. If you walked across country between these two points, would the distance be more or less than that covered by flying between the points? Why?

_____

_____

4. What is the bearing and distance from Seven Fountains road junction to where the trail crosses Shawl Gap?

5. What would the scale of this map have to be to show the same area on a sheet of paper one-fourth as large?

_____

6. What would the scale of this map have to be if one-fourth of the area shown had to cover the whole sheet?

_____

7. Construct a graphic scale for the answers to questions 5 and 6. Use the "parallel line" method and show your work.

8. Name two features shown on this map using dotted or dashed lines.

_____

_____

9. How can you tell that Blue Hole, on Passage Creek in the northeast portion of the map, is a water feature, other than by the blue color? (HINT: The answer is found on page 54.)

_____

_____

A-30

NAME _____

LAB SEC. _____

1. Contour Map A below, using a contour interval (C.I.) of 20 feet. After completing the map, using a 20 ft. C.I., re-contour the area to the southwest of Red Creek using a 5 ft. C.I. The resulting mess will illustrate the importance of a wise selection of contour interval. Draw contour lines with a PENCIL; do **not** use a ballpoint pen.

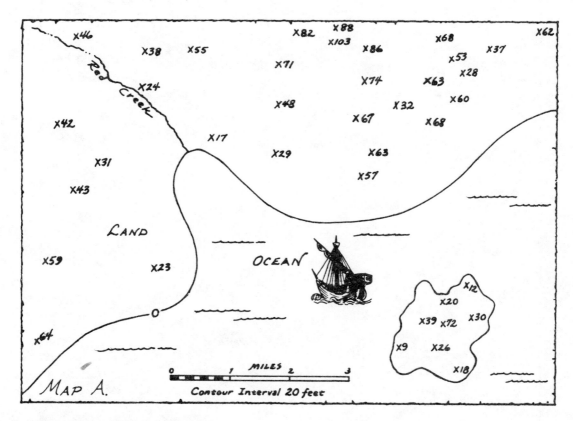

2. What is the relief of this map?

_____

3. What feature is represented by the close spacing of the contour lines in the SW corner of the map?

_____

4. Is the approximate center of the northeast quarter of this map a good place to build a factory? Explain.

_____

_____

5. What is the gradient in feet per mile from the highest point on the map to the mouth of the creek?

_____

6. What is the gradient from point 64 in the SW corner to the nearest shoreline?

_____

7. Contour Map B using a C.I. = 10 ft. Number the contour lines. Note that the 10 ft. contour line has been drawn in for you. Sketch in the stream drainage. Calculate the gradient of the longest stream from the highest point to the lowest point as you have contoured it.

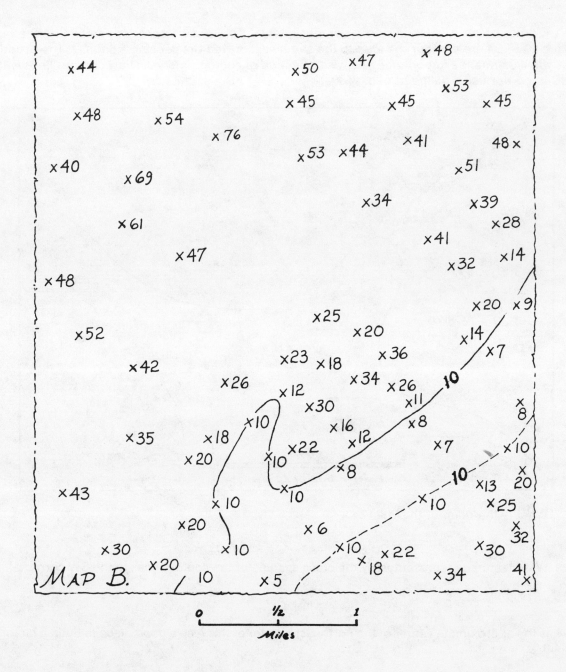

Gradient of the longest stream valley: _____ ft./mi.

UPHILL END

250

200

150

100

CONTOUR INTERVAL = 50 ft.

DOWNHILL END

FOLD UP — BOTH SIDES OF STREAM

STREAM FLOW

FOLD DOWN ON THIS SIDE OF THE DASHED LINE

LINE 2a

FOLD DOWN ON THIS SIDE OF THE DASHED LINE

LINE 2b

STREAM IS LINE 1

Carefully pull page out of manual and follow instructions below.

(1) Fold as indicated along crease lines 1, 2a and 2b.
(2) Hold at eye level with stream flow arrows pointing towards you, and tilt until each contour line appears level.

(3) Which way do the "V's" point? Upstream or Downstream?

EXERCISE 11, Continued

NAME   LAB SEC.

9. Construct a contour map in the space below according to the following:

   a. Area is an island, steeper on the north side than on the south side.
   b. Highest point is 86 feet above sea level.
   c. Contour interval is 10 feet.
   d. Horizontal scale is 1 inch represents 1 mile.
   e. The south slope should appear as a smooth slope with a gradient of 40 feet per mile.
   f. Two streams drain the island; one flows SE, the other almost due W. (HINT: Don't forget contour rule 8.)
   g. On the island are a house, a cemetery, and a windmill; show them by appropriate symbols.

ANTELOPE PEAK, ARIZ.
(Turn to map on page 111.)

1. Using the tier and range method, locate, to the nearest 40 acres, the overpass just north of the word "Antelope" in Antelope Wash.

_____

2. What is the elevation of the southwest corner of Sec. 3, R2E, T7S?

_____

3. Contrast the topography of this section to that of Sec. 34 just to the north. Does one section have rougher topography? If so, which one?

_____

A-35

MENAN BUTTES, IDA.

(Turn to map on page 41.)

1. What advantage does this map have over an ordinary contour map?

_____

_____

2. Examine the SE side of North Menan Butte. What disadvantages are there in using contour lines and shaded relief on the same map?

_____

_____

3. The map shading is due to light coming from which direction?

_____

BRIGHT ANGEL, ARIZ.

(Turn to map on page 45.)

1. Compare this map to the Cookeville East map. Which covers the largest area?

_____

Which has the greatest relief?_____

What is the maximum relief of the Bright Angel map?_____

2. Briefly describe the topography in the SW¼ of the map. (Is the land surface rough or smooth? Are the slope changes gradual or abrupt? Stairstepped?)

_____

_____

3. Which "rim" of the canyon is generally higher, north or south? _____

4. What is the contour interval of the map?_____

Would it be reasonable to use an interval of 20 ft.? _____

Explain your answer. _____

_____

_____

5. Would shading improve this map? Explain.

_____

_____

Compare the topographic maps in this exercise to any available maps using color layering or raised relief as a means of showing topography. Assure yourself of the relative accuracy of each type map and then evaluate the advantages and disadvantages of each for yourself.

EXERCISE 12:
TOPOGRAPHIC PROFILES

NAME   LAB SEC.

Equipment: Triangle, scale, **sharp** 3H pencils, graph paper (20 squares per inch). Suggestion: Get together with one or two of your classmates and buy a pack of paper. It's cheaper that way!

COOKEVILLE EAST, TENN.

(Turn to map on page 43.)

1. Construct the following topographic profiles:

Draw a profile along the line A-A' as shown on the map. Construct the profile first with a vertical exaggeration of 4X. Then using the same points, draw the same profile with no vertical exaggeration, then with a 10X vertical exaggeration, noting each time the change in overall shape of the profile.

BRIGHT ANGEL, ARIZ.

(Turn to map on page 45.)

1. Draw a profile along the line A-A' as shown on the map. Use a vertical scale of 1'' = 5000 ft. and calculate the vertical exaggeration. Are the slopes smooth and even? On your profile note how crossing Bright Angel Creek at an angle tends to make the stream appear wider and its bordering slopes smoother than the crossing of the Colorado River. Make sure you understand why.

HARRISBURG QUADRANGLE, PA.
(Turn to map on page 66)

1. Draw a profile along the line A-A' as shown on the map. Use a 2.6X vertical exaggeration. Is the relief sharp or gentle? Note that the use of "heavy" contour lines only is sufficient for the southeast side of Peters Mountain but not for the Clark Creek Valley. Do similar situations exist for the other mountains and valleys along the profile? Would the use of a 5X exaggeration improve the appearance or understanding of this profile?

2. Draw a profile along the line B-B' as shown on the map. Use a 2.6X vertical exaggeration. Compare this profile with the A-A' profile above. Which gives the truest picture of the ruggedness of Peters Mountain? Can you now explain the route of the highway that crosses Peters Mountain near the airway beacon?

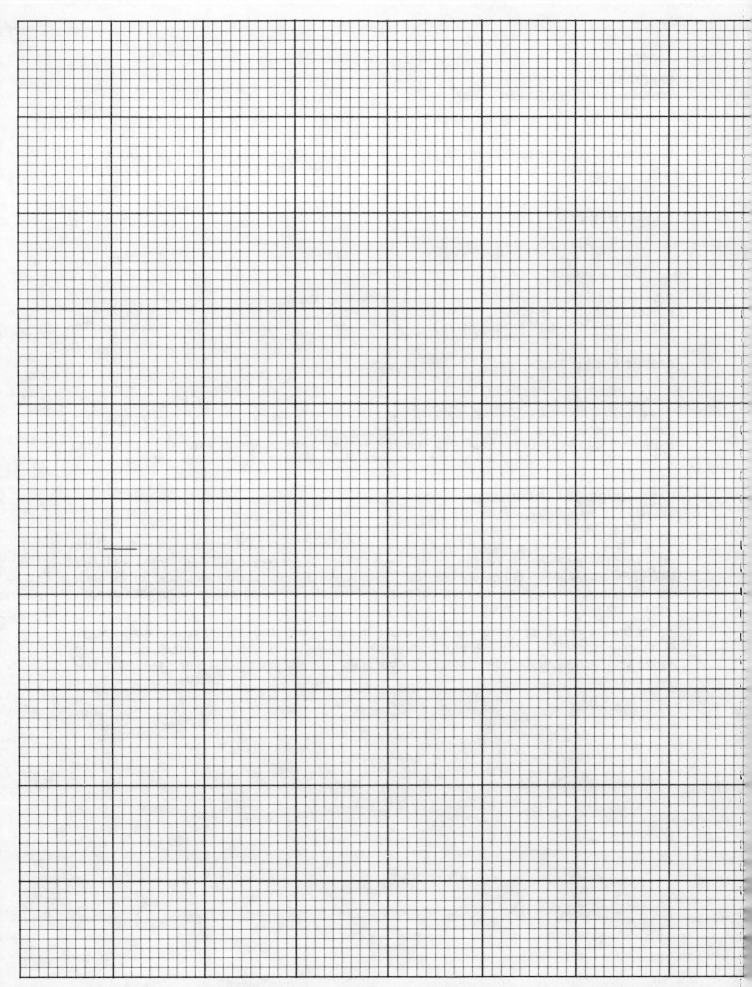

A-38

EXERCISE 13:
STREAMS AND STREAM SYSTEMS

NAME _____   LAB SEC. _____

GOVERNMENT SPRINGS, COLO.

1. Stream gradient of Happy Creek:_____

2. Topographic profile

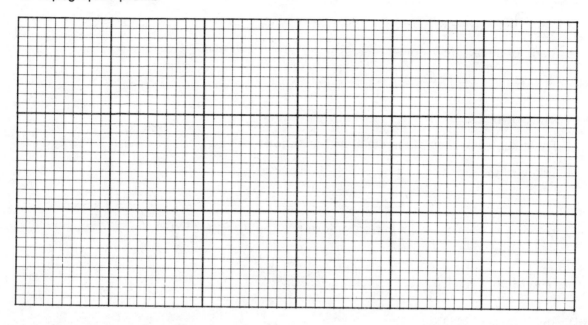

3. Transverse profile of stream valleys:_____

4. Flood plain features in the Government Springs area: _____

5. Stage of stream development:_____

6. Stage of topographic development:_____

LEAVENWORTH, KANS.-MO.

1. **Sketch** of profile across the
   Missouri River valley:
   (Use space to right.)

2. Shape of stream valley: _____

3. Sherman Army Airfield is built on the flat area known as _____

4. Origin of Mud Lake: _____

_____

_____

5. Stage of stream development represented by the Missouri River:

_____

CAMPTI, LA.

    1. Shape of the Red River stream valley:_____

_____

    2. DO ON MAP. LAB INSTRUCTOR MAY EXAMINE.

    3. The brown stippled pattern represents sand deposits known as

_____

    4. Estimated maximum gradient of the Red River:_____

    5. Stage of stream development represented by the Red River:

_____

BRETON SOUND, LA.

    1. The deltaic branches of the main river are known as _____

    2. Origin of the strips of land bordering the "passes":_____

_____

    3. Which way do the coastal currents move near South Pass? Explain.

_____

    4. Mississippi delta type:_____

ANDERSON MESA, COLO.

    1. Evidence of rejuvenation:_____

    2. Gradient of the Dolores River:_____

    3. Is this typical of meandering streams?_____

HOLLOW SPRINGS, TENN.

    1. Drainage pattern: _____

    2. Stage of topographic development, NW area:_____

    3. Stage of topographic development, SE area:_____

PLUM GROVE, TENN.-VA.

    1. Drainage pattern of Stanley Valley area: _____

    2. Type of structure probably responsible for rectangular drainage:

_____

    3. General description of rock units underlying this map area:

_____

EXERCISE 13, Continued

MAVERICK SPRING, WYO.

    1. Evidence for semi-arid climate:_____

_____

    2. Drainage pattern:_____

    3. General description of rock units underlying this map area:

_____

    4. Type of geologic structure:_____

MT. RAINIER, WASH.

    1. Drainage pattern:_____

STRASBURG, VA.

    1. What geologic explanation can you give for the "false meanders" shown by the North and South Forks of the Shenandoah River?

_____

_____

HARRISBURG, PA.

    1. Which is older, the Susquehanna River or the topographic ridge called Blue Mountain? Explain the reasoning behind your answer.

_____

_____

_____

    2. The Susquehanna River is an example of what sort of stream course?

_____

EXERCISE 14:
KARST AND GROUNDWATER

NAME   LAB SEC.

MAMMOTH CAVE, KY.

   1. What type of bedrock underlies the sinkhole plain? _____

   2. What type of bedrock underlies the southeastern corner of the map area and why are there no sinkholes in this area?

_____

   3. How many miles (minimum) does the Little Sinking Creek water travel underground to reappear at the springs at Turnhole Bend?

_____

   4. What is the pattern formed by the sinkholes in the solution valleys, and why?

_____

_____

LOST RIVER AREA, INDIANA

   1. Which of the two levels of Lost River has the higher (steeper) stream gradient, the underground or the surface level?

_____

INTERLACHEN, FLA.

   1. Why is Mirror Lake shaped the way it is? _____

_____

   2. How deep must you drill from your housesite to hit the water table? Do you sign a contract with this driller or look for another?

_____

   3. DO ON MAP. LAB INSTRUCTOR MAY EXAMINE.

   4. Is the water table in this area flat or sloping? _____
If sloping, which way?

_____

BOTTOMLESS LAKES, N. MEX.

   1. Why the name ''Artesia''? _____

_____

   2. Discuss the present and past stream courses of the small intermittent stream that presently flows into Dimmitt Lake.

_____

_____

_____

NAME _____   LAB SEC. _____

ENNIS, MONT.

1. DO ON MAP. LAB INSTRUCTOR MAY EXAMINE.

2. What stage of topographic development is represented by this area?

_____

FURNACE CREEK, CALIF.

1. Why is the lake shown with ruled lines instead of solid blue?

_____

2. What is the geologic name for this type of lake? _____

3. What evidence is there that this area is a depression?

_____

4. What stage of topographic development is represented by this area?

_____

ANTELOPE PEAK, ARIZ.

1. What stage of topographic development is represented by this area?

_____

2. Vertical relief from the SW corner of section 34 to the SW corner of section 5:

_____

3. Slope of the pediment/bajada surface, in degrees: _____

4. Explain the crinkly nature of the contour lines.

_____

_____

_____

_____

NEW HOME, TEX.

1. List of map features suggesting arid climate (list at least 5):

_____

_____

_____

_____

_____

2. Origin of the "buffalo wallows" _____

_____

GLENROCK, WYO.

1. What type of dunes are shown in the NE corner of section 23?

_____

2. **From** which direction does the prevailing wind blow? (It is standard procedure to designate winds according to the direction from which they blow; e.g., the north wind blows from the north towards the south.)

_____

ASHBY, NEB.

1. Are these transverse or longitudinal dunes? _____

2. **From** which way did the prevailing wind blow? _____

3. Cite evidence that the dunes are stabilized. _____

_____

_____

NAME _____ LAB SEC. _____

McCARTHY, ALASKA

1. The rock train down the center of Nizina Glacier is a _____

_____

2. Evidence that Chitistone Glacier formerly extended down the full length of Chitistone River valley:

_____

_____

_____

3. Are the glaciers advancing or retreating? Cite more than one line of evidence.

_____

_____

_____

CHIEF MOUNTAIN, MONT.

1. Mt. Merritt, Ipasha Peak, Ahern Peak, and Mt. Wilbur are all:

_____

2. The rock basin just north of Mt. Wilbur is an example of a:

_____

3. Iceberg Lake is a type of meltwater lake known as a:

_____

4. The chain of lakes below Ipasha Glacier are called _____ lakes.

5. Has Chief Mountain been glaciated? Explain your answer.

_____

_____

HOLY CROSS, COLO.

1. Sketch a cross section of Lake Fork valley. Is your sketch U-shaped or V-shaped?

_____

2. Busk Creek valley is what type of glacial valley?

_____

3. What is the origin of the natural dam that created Turquoise Lake? Name the topographic feature.

_____

4. The waters of Homestake Creek eventually flow into:

_____

HAYDEN PEAK, UTAH-WYO.

1. The narrow ridge of horns and cols is a feature of glacial erosion known as:

_____

2. What evidence can you see that suggests that alpine glaciation in this area ended quite some time ago?

_____

_____

MT. RAINIER, WASH.

1. "Glacier Island" and "Echo Rock" are examples of:_____

2. Name of an arête:_____

EXERCISE 17:
CONTINENTAL GLACIATION

NAME _____ LAB SEC. _____

KAATERSKILL, N.Y.

1. What evidence can you see that Cooper Lake used to drain to the west?

_____

_____

_____

2. Will future stream piracy affect the headwaters of Beaver Kill? Explain.

_____

_____

_____

PALMYRA, N.Y.

1. These streamlined hills are known as:_____

2. Longitudinal profile:

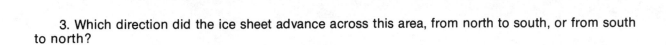

3. Which direction did the ice sheet advance across this area, from north to south, or from south to north?

_____

4. Which direction did the ice **flow** as the ice sheet retreated?

_____

JACKSON, MICH.

1. What is the name and origin of the deposit that covers most of the Jackson map?

_____

2. How may the drainage in this area be described? _____

3. What is the geologic name for a feature like Blue Ridge? Explain its origin.

_____

_____

WHITEWATER, WIS.

    1. DO ON MAP. LAB INSTRUCTOR MAY EXAMINE.

    2. Which direction did the ice sheet advance?

_____

BOOTHBAY, MAINE

    1. Is the topography here the result of glacial erosion or glacial deposition? Explain your reasoning.

_____

_____

_____

_____

NAME _____   LAB SEC. _____

BOOTHBAY, MAINE

1. Is this a shoreline of submergence or emergence? _____

2. List 3 features that support the above answer.

Geologic Feature	**Map Name**
_____	_____
_____	_____
_____	_____

3. Possible explanation for the steepsidedness of channels as Back River and Robinhood Cove.

_____

_____

_____

4. Does your answer 3 also account for the parallelism of the ridges and valleys? Explain.

_____

_____

5. Explain the absence of depositional features in this map area.

_____

_____

6. Is the water here shallow or relatively deep for near shore water? How can you tell?

_____

_____

7. Give the map name of an example of each of the following features:

   a. bay _____

   b. estuary _____

   c. cove _____

   d. rocks hazardous to navigation _____

8. Are good harbors plentiful or scarce? _____

9. Is this area safe or dangerous for incoming ships on a foggy night?

_____

_____

LYNN, MASS.

1. Is this a shoreline of submergence or emergence. List three map features that support your answer.

_____   _____

_____

2. Where is a tombolo shown on the map?_____

   Is it a complex or a simple tombolo?_____

3. Name a bayhead beach._____

4. What is the nature of "Johns Peril"?_____

5. From which direction does the strongest wave action come? Cite your evidence.

_____

_____

6. Why is the water shallower on one side of Little Nahant Island than on the other?

_____

_____

7. What do the dashed blue lines indicate? _____

_____

POINT REYES, CALIF.

1. What is the map scale?_____

   What is the contour interval?_____

2. Is the relief greatest in the eastern or western half of the map?

_____

3. What is likely to happen in the future to Drakes Estero and Estero de Limantour? _____

_____

_____

4. Give the map name of an example of each of the following features:

   a. spit   _____

   b. cliffed headland   _____

   c. stack   _____

   d. cove   _____

   e. estuary   _____

   f. beach   _____

5. Where is the ocean floor steeper: in Drakes Bay, or just south of Point Reyes?

_____

6. Where would the best harbor be in this map area? Explain your reasoning.

_____

_____

_____

7. Is there a longshore current operating in Drakes Bay? If so, which way does it flow?

_____

BEAUFORT, N.C.

1. What is the map scale? _____

   What is the contour interval? _____

2. Is this a shoreline of submergence or emergence? Explain your reasoning.

_____

_____

_____

3. Give the map name of an example of each of the following features:

   a. lagoon _____

   b. barrier bar _____

   c. estuary _____

   d. inlet _____

4. What does the future hold for Core Sound? _____

_____

_____

5. Is this an area of high or low relief? _____

6. Do the streams in this area have high or low gradients? _____

7. Does the water in Adams Creek Canal flow the same direction all day long? Why not?

_____

_____

SAN CLEMENTE ISLAND CENTRAL, CALIF.

1. Construct a topographic profile through A-A'. Use a vertical exaggeration of 2X. Extend the southwest end of the profile beneath the water and show what you think the bottom profile might be.

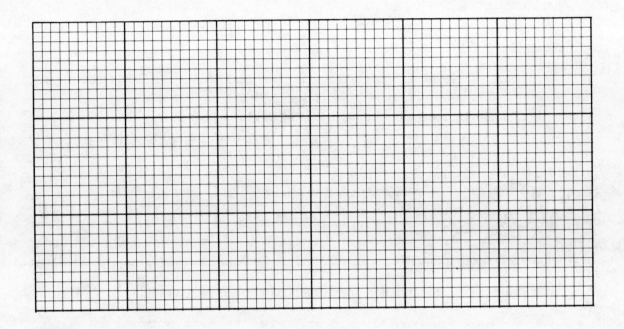

2. What is the origin of the alternating steep, cliff-like slopes and relatively flat terraces which show on both the map and topographic profile?

_____

_____

_____

3. Do these map features indicate that there has been a relative change in sea level in this area? If so, did sea level rise or fall? How much?

_____

_____

EXERCISE 19:
VOLCANIC LANDFORMS

NAME   LAB SEC.

MENAN BUTTES, IDA.

1. Topographic profile of North Menan Butte:

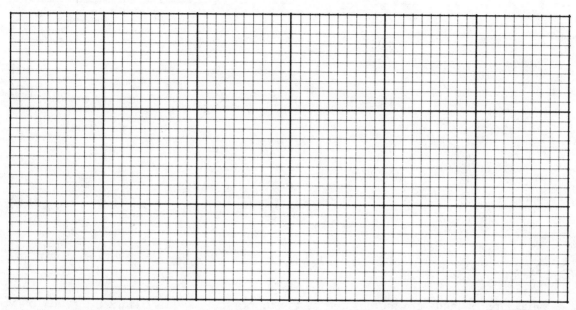

    2. What is the vertical relief from the highest peak of North Menan Butte down to the Henry's Fork of the Snake River?

_____

    3. What type of volcanic cones are the Menan Buttes? _____

MT. RAINIER, WASH.

1. What evidence can you see on the map that Mt. Rainier is a volcano?

_____

2. Does Mt. Rainier appear to be active or inactive? Explain your answer.

_____

3. What is the vertical relief from the peak down to the Superintendent's Headquarters?

_____

4. What type of volcanic cone is Mt. Rainier? _____

MAUNA LOA, HAWAII

1. What is the most recent volcanic activity shown on this map?

_____

    2. What is the vertical relief from the highest point down to the lowest index contour in the SE corner of the map?

_____

3. What type of volcanic cone is Mauna Loa? _____

MT. DOME, CALIF.-ORE.

1. Topographic profile across the three scarps:

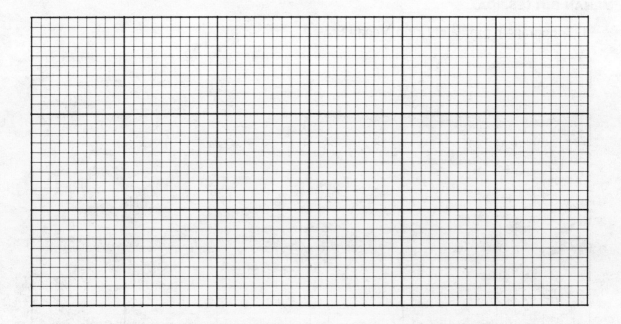

2. What type of faults are represented by the topographic scarps? **Mark the faults on your profile and indicate with arrows the direction of movement suggested by the topography.** Type of faults:

_____

3. DO ON MAP. LAB INSTRUCTOR MAY EXAMINE.

4. Were the fault scarps already there at the time of the eruption, or is the faulting post-eruption? Explain your answer.

_____

_____

_____

CRATER LAKE NATIONAL PARK, ORE.

1. What is the greatest water depth shown on the map?_____

2. How deep is the water over Merriam Cone? _____

SHIP ROCK, N. MEX.

1. DO ON MAP. LAB INSTRUCTOR MAY EXAMINE.

2. What is the proper geologic name of the ridges that radiate out from the volcanic neck, and how did they form?

_____

_____

EXERCISE 20:
MAN AND HIS ENVIRONMENT

NAME _____ LAB SEC. _____

AJO, ARIZ.

1. How deep is the New Cornelia Mine pit? _____

2. How wide is the New Cornelia Mine pit? _____

3. Estimate the combined acreage of the North, South, and East tailings ponds.

_____

4. Where does the water for the copper mill come from and what might be the long range consequences of using such large quantities of water?

_____

_____

_____

McCARTHY, ALASKA

1. List some of the logistical problems associated with mining in remote places, particularly places with such harsh climates as Alaska.

_____

_____

_____

2. What are some of the problems that might be encountered while building roads, landing strips, etc., in this area?

_____

_____

_____

HARRIMAN, TENN.

1. How much acreage has been utilized thus far in the ash disposal area?

_____

2. Where does the water used in the steam plant come from, and what are the potential problems involved in returning it to its source?

_____

_____

_____

RIVERTON, MINN.

    1. List several ways in which the local inhabitants might benefit by iron ore mining in this area.

_____

_____

    2. List several ways in which iron ore mining might cause environmental problems in this area.

_____

_____

_____

    3. What environmental problems might result if an old mine pit is used as a garbage dump?

_____

_____

_____

_____

OSWEGO, KANS.

    1. What is the total tonnage of coal produced from an entire section if the coal seam is 3 feet thick and 1800 tons are produced per acre foot?

_____

    2. If the profits were $5.00 per ton, what would the profit be from a complete section.

_____

    3. If the cost of land restoration is $1000 per acre, what would be the total cost of restoration for a complete section? Is this expenditure justifiable, and are you, as a consumer, willing to pay for it in terms of higher coal (electricity) prices?

_____

_____

FAIRBORN, OHIO

    1. What will happen to Wright-Patterson Air Force Base if the flood gates of Huffman Dam are closed?

_____

    2. In what way could deliberately inundating the base and part of the town of Fairborn serve as a flood control measure? How could this drastic measure be justified?

_____

_____

_____

DEROUEN, LA.

    1. Why do you think oil and gas might accumulate in the sedimentary layers around the flanks of salt domes? What would allow the oil and gas to be trapped here?

_____

_____

_____

EXERCISE 21:
GEOLOGIC MAPS AND CROSS SECTIONS

NAME _____   LAB SEC. _____

FACTORS GOVERNING THE WIDTH OF OUTCROP PATTERNS ON GEOLOGIC MAPS

The astute student will have noticed while studying the block diagrams in Fig. 21-1 that the **width** of the outcrop of a particular formation sometimes exceeds its true thickness (true thickness is measured from the top to the bottom of a formation, perpendicular to its upper and lower contacts). The width of the outcrop of a rock unit as shown on a geologic map varies according to two factors:

1. the thickness of the rock unit;

2. the angle at which the topographic surface cuts across the rock unit.

Several diagrams below show how the outcrop width of a formation will vary with the aforementioned factors. Nine cross sections are shown, along with three geologic sketch maps. There are six incomplete geologic sketch maps, which you are to complete.

SKETCH IN THE SIX INCOMPLETE GEOLOGIC MAPS. USE THE SAME ROCK TYPE SYMBOLS ON YOUR MAPS AS USED IN THE CROSS SECTIONS, OR, IF YOU PREFER, USE DIFFERENT COLORS FOR THE DIFFERENT ROCK TYPES.

Factor #1: THICKNESS of the FORMATION:

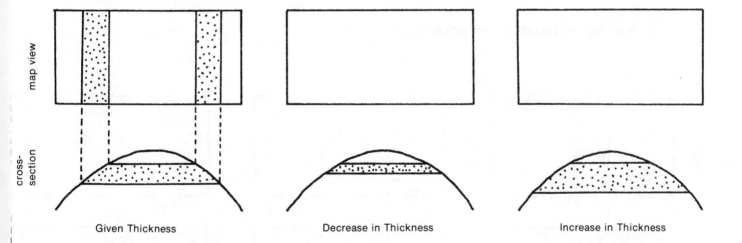

Given Thickness   Decrease in Thickness   Increase in Thickness

Factor #2: the angle at which the topographic surface cuts across the formation depends on a) the slope of the topography, and b) the attitude (strike & dip) of the layered unit. The diagrams on the following page illustrate these relationships.

In the meantime, you should note that **these rules apply only to layered rock bodies.** In the case of irregular intrusive bodies, the width of the outcrops as shown on geologic maps cannot be readily predicted. It is equally difficult to predict the underground or cross-sectional configuration of these intrusions from known outcrop patterns.

Factor #2a: TOPOGRAPHIC SLOPE:

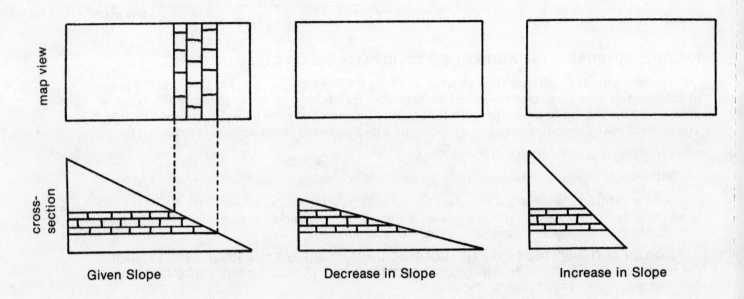

map view

cross-section

Given Slope   Decrease in Slope   Increase in Slope

General rule: the steeper the slopes, the narrower the outcrops appear on the map.

Factor #2b: ATTITUDE of the ROCK UNIT:

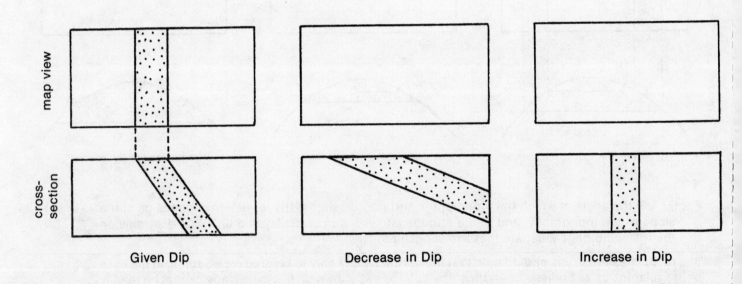

map view

cross-section

Given Dip   Decrease in Dip   Increase in Dip

Note that in the last example the rock layer is perpendicular to the topographic surface, and the outcrop width is the true thickness.

EXERCISE 21, Continued

SW-NE Topographic Profile and Geologic Cross Section of the Muddy Creek area, Flagstaff Peak Quad, Utah

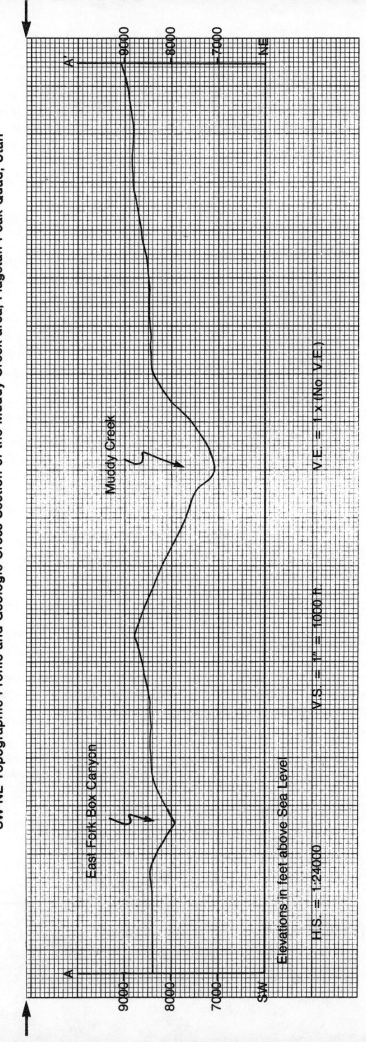

Elevations in feet above Sea Level

H.S. = 1:24000   V.S. = 1" = 1000 ft   V.E. = 1 x (No V.E.)

Instructions: Remove this page from the manual and fold it over at the arrows so that you can place the edge of the graph paper along the profile line A-A' on the Flagstaff Peak geologic quad. **Mark the geologic contacts** on the edge of the graph paper, in the same manner that you learned to mark the contour lines when taking data points for constructing a topographic profile. Next, **mark these contacts on the topographic profile** itself. Now draw a line from the Kc/Kb contact on the northeast side of East Fork Box Canyon through the profile to the Kc/Kb contact on the southwest side of the valley of Muddy Creek. This line represents the base of the Castlegate Sandstone (Kc), and also the top of the Blackhawk Formation (Kb). It is a straight line

because in this simple "layer cake" geology the beds are nearly flat-lying layers. Now draw in the Kc/Kb contact line on the other side of the Muddy Creek valley, and proceed to draw in all the other contacts similarly. **When you have all the contacts drawn in, color in the different formations** and you will have constructed the geologic cross section for this profile.

Note that "layer cake" geology is the simplest of all to draw in cross section. Later you will be asked to draw a cross section in an area of folded rock layers, and your ability to envision the subsurface configuration of the layers will receive a more challenging test.

WETTERHORN PEAK, COLO.

1. What type of igneous unit is the quartz latite, intrusive or extrusive? Explain your answer.

_____

_____

_____

2. What is the proper geologic term used to describe long, narrow igneous bodies such as those radiating out from the main quartz latite bodies?

_____

3. Are the San Juan and Potosi Formations intrusive or extrusive? Explain the logic behind your answer.

_____

_____

_____

4. List these units in order of their ages, from oldest to youngest: Tql, Tsj, Tp. List them in a vertical column, with the oldest on the bottom.

_____

_____

_____

5. What type of geologic structure might account for the rectangular outlines of the main quartz latite body and the associated dike orientations?

_____

6. What is the dip of the fault plane in the NE corner of the map?

_____

7. What geologic process is responsible for the semicircle of cliffs surrounding the talus at "C"?

_____

MIFFLINTOWN, PA.

1. Are the rock layers in this area flat-lying, tilted, or folded? Cite at least two lines of evidence to support your answer.

_____

_____

2. Does B-B' cross any anticlines or synclines? How many of each (if any)?

_____

3. DO ON MAP. LAB INSTRUCTOR MAY EXAMINE.

EXERCISE 21, Continued

NW-SE Topographic Profile and Geologic Cross Section of the New Lancaster Valley area, Mifflintown, Quad, Pa.

Jacks Mtn.

High Top

New Lancaster Valley
Honey Creek

B

B'

Sea Level

NW

SE

2000

1000

Sea Level

1000

2000

H.S. = 1:24000   V.S. = 1" = 2000 ft.   V.E. = 1 x (No V.E.)

Instructions: Remove this page from the manual and fold it over at the arrows so that you can place the edge of the graph paper along the profile line B-B' on the Mifflintown geologic quad. **Mark the geologic contacts** on the edge of the graph paper, in the same manner that you learned to mark the contour lines when taking data points for constructing a topographic profile. Next, **mark these contacts on the topographic profile** itself. Extending these contacts underground through the profile to construct the geologic cross section will be more complicated for this section than for the "layer cake" section you drew earlier. The rock units here have been folded into an anticline and a syncline, and the contact lines you draw must be curved appropriately. You may start with the contacts nearest the

strike and dip symbols: **project the contacts downwards at the angle of the dip, then gradually curve them around through the folds** as dictated by the nature of the structures themselves. A contact which starts down on one side of the syncline should reemerge at the surface on the opposite side of the fold. **Important:** the thickness of each formation should remain constant throughout the cross section.

Note: When the rock units are folded, faulted, tilted or anything other than flat-lying it is important that no vertical exaggeration be used in the construction of a geologic cross section, for vertical exaggeration will distort the dip angles, fold shapes, etc.

A-63

DANFORTH, MAINE

    1. Is the large fold structure in this area a syncline or an anticline? _____

How can you determine this from the outcrop pattern?_____

_____

What structural symbols tell you what the fold is?_____

    2. Explain two ways you could determine from the geologic map that the Devonian granite is younger than the Ordovician and Silurian rocks even if the rock units had no symbols indicating their ages.

_____

_____

_____

    3. Is the fault in the southwestern part of the map a strike-slip, a dip-slip, or an oblique-slip fault? (Oblique-slip means that the movement along the fault is neither horizontal nor in the dip direction, but at some in-between angle.) Explain the reasoning behind your answer choice.

_____

_____

_____

    4. What is the proper geologic term for a "horseback"?_____

    5. Name at least two other features shown on the map which are indicative of continental glaciation.

_____

_____

BRISBIN, MONT.

    1. What is the age of the oldest rock unit in this area? _____

Is it sedimentary, intrusive, extrusive, or metamorphic? _____

Is it exposed in a syncline or an anticline? _____

    2. What is the age of the youngest map unit on this map? _____

Is it a lithified rock unit or an unconsolidated sediment? _____

Where and by what was this unit deposited? _____

_____

EXERCISE 21, Continued

3. List the three periods not represented by rocks in this area, starting with the oldest at the bottom of the list.

_____

_____

_____

Why are there no rocks of these periods in this area? What might have prevented them from being deposited? Or, if they were deposited, what might have caused them to be removed?

_____

_____

_____

What is the geologic term used to describe a contact such as that at "U"?

_____

4. Is the east-west trending fault marked "T" a thrust fault or a high angle fault?

_____

What kind of fault is South Dry Creek Fault? _____

Which of the faults is older? Explain your reasoning. _____

_____

Which way does the thrust fault plane dip: north, south, east, or west? _____

Which way did the overthrust block move: north, south, east, or west? _____

5. In the 12 lines provided below, reconstruct the geologic history of this area, starting with the first event on the bottom line, finishing with the most recent event in the top line. (The 12 geologic events are listed in scrambled order on the page facing the Brisbin map.)

_____

_____

_____

_____

_____

_____

_____

_____

_____

_____

_____

_____